Operation-Based Infinite-Queue
SBC Process Algebra
For Systems Modeling

-- General Systems Theory 2.0 at Work --

William S. Chao

Structure-Behavior Coalescence

$$\text{Systems Architecture} = \text{Systems Structure} + \text{Systems Behavior}$$

CONTENTS

PREFACE

The need for systems modeling arises because any real-life system is inherently complicated. It is impossible to comprehend fully the intricate interaction of any system of the real world with its environment, or to define all its components and each of its details. Systems modeling or system modeling is an artifact created by humans to define what a system is.

Process algebras are a diverse family of related approaches to the study of concurrent systems. Their tools are algebraic languages for the high-level description of interactions, communications and synchronizations between a collection of independent agents or processes. Process algebras also provide algebraic laws that allow process descriptions to be manipulated and analyzed, and permit formal reasoning about equivalences and observation congruence among processes. Accordingly, process algebra provides a perfect method for system modeling.

Operation-based infinite-queue SBC process algebra (O-I-SBC-PA) is one of the six specialized SBC process algebras. In this book, we use O-I-SBC-PA to achieve the robust systems modeling of a system. To see is to believe. Therefore, many examples are presented to help the reader fully understand the use of O-I-SBC-PA.

ABOUT THE AUTHOR

Dr. William S. Chao is the CEO & founder of SBC Architecture International®. SBC (Structure-Behavior Coalescence) architecture is a systems architecture which demands the integration of systems structure and systems behavior of a system. SBC architecture applies to hardware architecture, software architecture, enterprise architecture, knowledge architecture and thinking architecture. The core theme of SBC architecture is: "Architecture = Structure -->> Behavior."

William S. Chao received his bachelor degree (1976) in telecommunication engineering and master degree (1981) in information engineering, both from the National Chiao-Tung University, Taiwan. From 1976 till 1983, he worked as an engineer at Chung-Hwa Telecommunication Company, Taiwan.

William S. Chao received his master degree (1985) in information science and Ph.D. degree (1988) in information science, both from the University of Alabama at Birmingham, USA. From 1988 till 1991, he worked as a computer scientist at GE Research and Development Center, Schenectady, New York, USA.

Dr. William S. Chao has been teaching at National Sun Yat-Sen University, Taiwan since 1992 and now serves as the president of Association of Enterprise Architects, Taiwan Chapter. His research covers: systems architecture, hardware architecture, software architecture, enterprise architecture, knowledge architecture and thinking architecture.

PART I: BASIC CONCEPTS

Chapter 1: Introduction to Systems Modeling

All things that amaze us as something independent are essentially parts of a system. We usually call the parts of a system its components. Every system is something the whole. Systems emphasize the holistic vision.

The need for systems modeling arises because any real-life system is inherently complicated. It is impossible to comprehend fully the intricate interaction of any system of the real world with its environment, or to define all its components and each of its details. Systems modeling or system modeling is an artifact created by humans to define what a system is.

In this chapter, we first introduce physical and conceptual systems. A physical system exists in the physical, concrete, or real world. A conceptual system exists in the conceptual, abstract, or virtual world. We then introduce the boundary of a system. A system has a boundary. The system itself is inside the boundary and the environment is outside the boundary. After these introductions, we shall thereafter discuss the systems modeling 1.0 as well as the systems modeling 2.0.

1-1 Physical and Virtual Systems

In general, systems are divided into two categories: 1) physical systems and 2) virtual systems.

A physical system exists in the physical world [Acko68]. A physical system is also called a concrete or real system. For example, a *telephone* composed of *microphone*, *earphone and keypad*, shown in Figure 1-1, is a physical, concrete, or real system.

Figure 1-1   A Telephone is a Physical System

As a second example, a *stool* composed of *seat* and *legs*, shown in Figure 1-2, is a physical, concrete, or real system.

Figure 1-2   A Stool is a Physical System

A virtual system is a system that is composed of non-physical components, i.e., ideas, thoughts, or notions. A virtual system exists in the virtual, abstract, or notional world. For example, the "*Snow White and the Seven Dwarfs*" fairy tale composed of "*Snow White*" and "*Seven Dwarfs*," shown in Figure 1-3, is a virtual, abstract, or notional system.

Figure 1-3   *Snow White and the Seven Dwarfs* is a Virtual System

As a second example, the *"Multi-Tier Personal Data System"* software composed of *MTPDS_GUI*, *Age_Logic*, *Overweight_Logic* and *Personal_Database*, shown in Figure 1-4, is a virtual, abstract, or notional system.

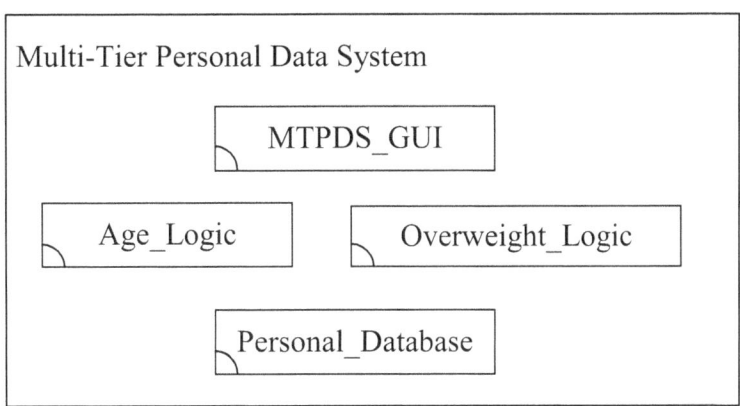

Figure 1-4   *Multi-Tier Personal Data System* is a Virtual System

1-2 Boundary and Environment of a System

We scope a system by describing its boundary as shown in Figure 1-5. All components of the system are inside the boundary while the environment is outside the boundary.

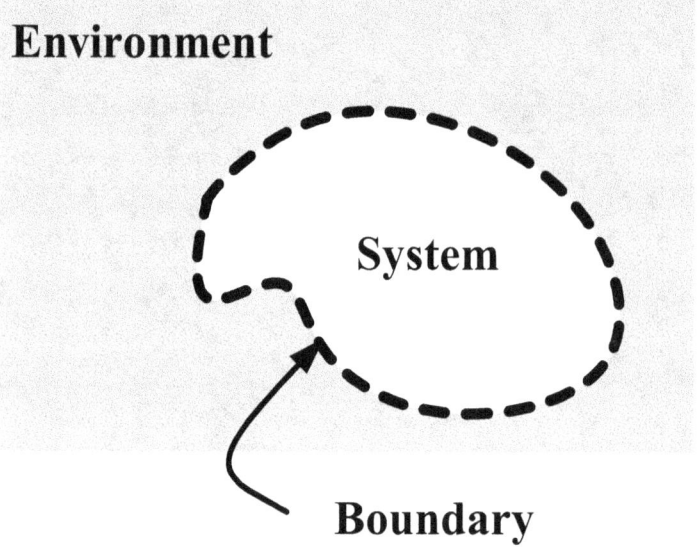

Figure 1-5   Boundary and Environment of a System

The environment is also known as the surroundings. A system may or may not interrelate with the environment. An open system interrelates with the environment through the exchange of matter, energy, data, information, or message as shown in Figure 1-6.

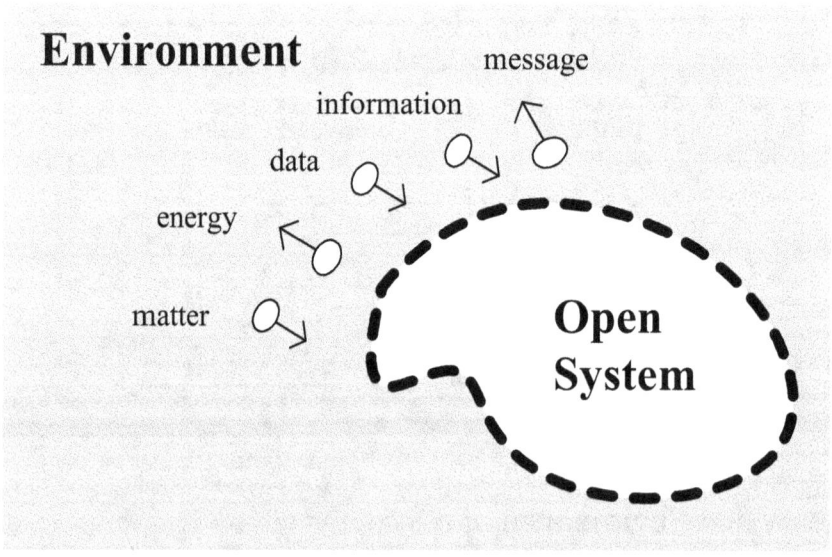

Figure 1-6   Open System Interrelates with the Environment

An isolated system does not interrelate with the environment at all. There is no exchange of matter, energy, data, information, or message between the isolated system and the environment as shown in Figure 1-7.

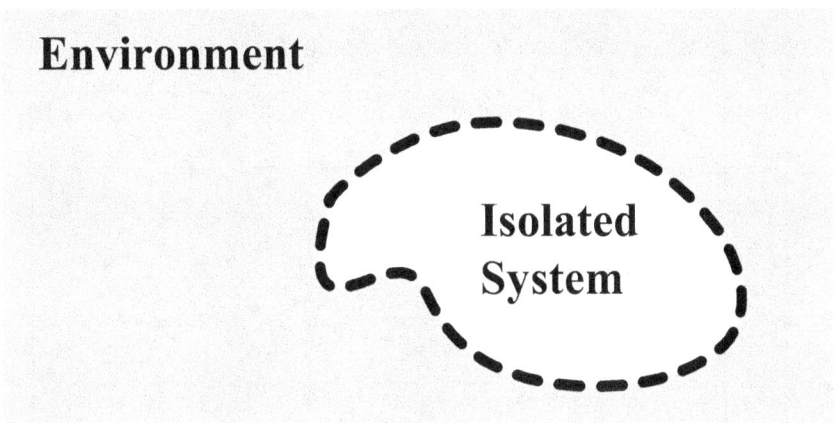

Figure 1-7   Isolated System Does Not Interrelate with the Environment

1-3 Systems Modeling 1.0

Systems modeling 1.0 defines a system, in Figure 1-8, hopefully to be an integrated whole, embodied in its assembled components, their interrelationships with each other and the environment [Chec99, Frie11, Ghar11, Mead08].

> A system, hopefully is an integrated whole,
> embodied in its assembled components,
> their interrelationships with each other and the environment.

Figure 1-8   Systems Modeling 1.0 Defining a System

Components are sometimes labeled as parts, entities, objects, building blocks and non-aggregated systems [Chao14a, Chao14b, Chao14c]. Interrelated components make a system not only a whole but also hopefully an integrated whole.

A system defined by systems modeling 1.0 has the following characteristics: 1) hopefully, it is an integrated whole; 2) it is embodied in its assembled components; 3) components are interrelated with each other and the environment; and 4) it uses structural decomposition [Chao14b, Ghar11] rather than functional decomposition [Scho10].

The structural decomposition method is to decompose a system into a number of components, as shown in Figure 1-9. Breaking down a large problem into a number of components to solve, is a relatively preferred method.

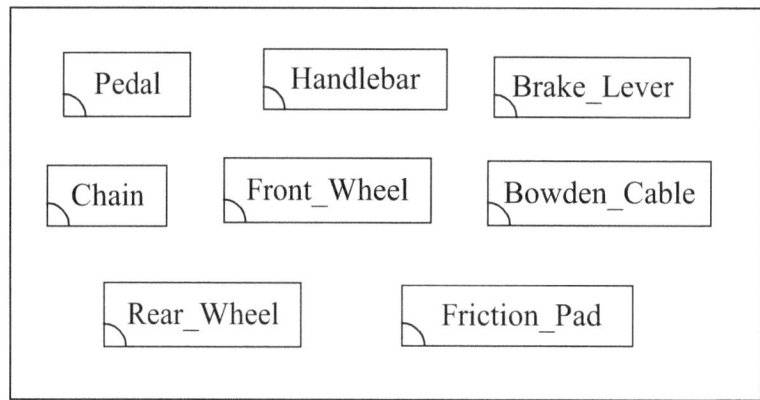

Figure 1-9   Structural Decomposition Method

The functional decomposition method is to decompose a system into a number of functions, as shown in Figure 1-10. Breaking down a large problem into a number of functions to solve, is a relatively non-preferred method.

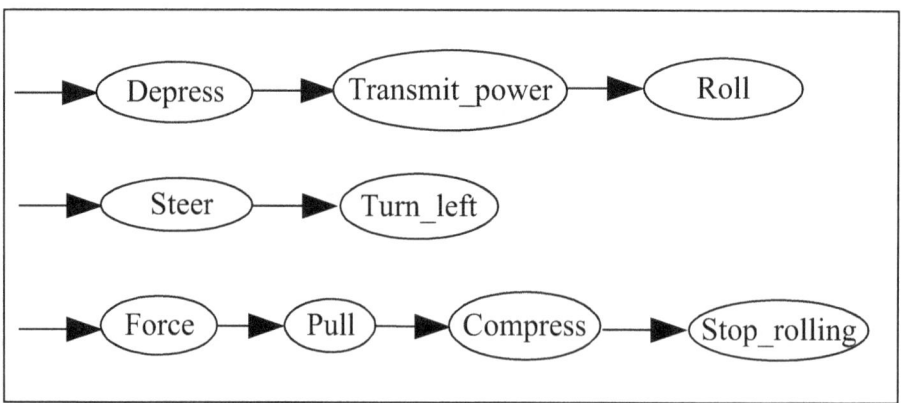

Figure 1-10   Functional Decomposition Method

1-4 Systems Modeling 2.0

Systems structure and systems behavior are the two most significant views of a system. In order to achieve a truly integrated whole of a system, we first need to integrate the systems structure and systems behavior together. In other words, integration of systems structure and systems behavior results in the integration of a whole system. Since systems modeling 1.0 does not define the integration of systems structure and systems behavior, very likely it only hopes and will never be able to really form an integrated whole of a system. In this situation, systems modeling 1.0 is powerless in defining a system appropriately.

Structure-behavior coalescence (SBC) provides an elegant way to integrate the systems structure and systems behavior of a system. A system is therefore redefined, by systems modeling 2.0, truly to be an integrated whole, through structure-behavior coalescence, embodied in its assembled components, their interactions with each other and the environment as shown in Figure 1-11.

A system,
through the SBC approach,
truly is an integrated whole,
embodied in its assembled components,
their interactions with each other and the environment.

Figure 1-11   Systems Modeling 2.0 Defining a System

Systems modeling or system modeling 2.0 formally defines the essence of a system and its details at the same time. Since systems modeling 2.0 demands the integration of systems structure and systems behavior, definitely it is able to form an integrated whole of a system. In this situation, systems modeling 2.0 is fully capable of defining a system.

A system defined by systems modeling 2.0 has the following characteristics: 1) it emphasizes the system's structure-behavior coalescence; 2) it is a truly integrated whole; 3) it is embodied in its assembled components; 4) components are interacting (or handshaking) with each other and the environment; and 5) it uses structural decomposition [Chao14a, Chao14b, Chao14c, Ghar11] rather than functional decomposition [Scho10].

Chapter 2: Introduction to Process Algebra

In this chapter, we first discuss what process algebra is. Then we show several examples of process algebras. After that we elaborate on the generalized SBC process algebra and six specialized SBC process algebras.

2-1 What is Process Algebra?

Process algebras are a diverse family of related approaches to the study of concurrent systems [Berg87, Chao15d, Hoar85, Miln89, Miln99]. Their tools are algebraic languages for the high-level description of interactions, communications and synchronizations between a collection of independent agents or processes.

Process algebras also provide algebraic laws that allow process descriptions to be manipulated and analyzed, and permit formal reasoning about equivalences and observation congruence among processes.

2-2 Examples of Process Algebras

There are several leading algebraic approaches to modeling concurrent systems.

Communicating Sequential Processes (CSP) [Hoar85] was first described in a 1978 paper by C. A. R. Hoare.

Arthur John Robin Gorell Milner introduced the Calculus of Communicating Systems (CCS) [Miln89, Miln99] around 1980.

Algebra of Communicating Processes (ACP) [Berg87] was initially developed by Jan Bergstra and Jan Willem Klop in 1982.

2-3 Generalized SBC Process Algebra

Generalized SBC process algebra (G-SBC-PA) evolved from CCS (Calculus of Communicating Systems) [Miln89, Miln99].

CCS is a general process algebra language for the study of communication and concurrency. Like CCS, generalized SBC process algebra is also a general process algebra language for the study of communication and concurrency.

2-4 Specialized SBC Process Algebras

Channel-based single-queue SBC process algebra (C-S-SBC-PA) [Chao15d, Chao15e, Chao15g], channel-based multi-queue SBC process algebra (C-M-SBC-PA) [Chao15d, Chao15f, Chao15h], channel-based infinite-queue SBC process algebra (C-I-SBC-PA) [Chao15b, Chao15c, Chao15d], operation-based single-queue SBC process algebra (O-S-SBC-PA), operation-based multi-queue SBC process algebra (O-M-SBC-PA) [Chao15d, Chao15f, Chao15h] and operation-based infinite-queue SBC process algebra (O-I-SBC-PA) [Chao15b, Chao15c, Chao15d] are the six specialized SBC process algebras.

Channel-based single-queue SBC process algebra, channel-based multi-queue SBC process algebra, channel-based infinite-queue SBC process algebra, operation-based single-queue SBC process algebra, operation-based multi-queue SBC process algebra and operation-based infinite-queue SBC process algebra all evolved from CCS (Calculus of Communicating Systems) [Miln89, Miln99].

CCS is a general process algebra language for the study of concurrent systems. Unlike CCS, six specialized SBC process algebras are only applicable to systems modeling [Burd10, Maie09, Chao16b, Chao16c, Chao16d, Chao16e, Chao16f, Chao16g, Craw15, Dam06, O'Rou03, Putm00, Rayn09, Roza11, Toga08].

Chapter 3: Mathematics of SBC Process Algebra

To give the SBC process a mathematical definition, we need a means to form new processes from old ones. The basic operators, always present in some form or other, allow sequentialization of prefixes or summation of processes or parallel composition of processes or recursive definition of a process or replication of processes or conditional definition of a process or null process.

3-1 Sequentialization of Prefixes

Sometimes prefixes must be temporally ordered. For example, it might be desirable to specify algorithms such as: execute the "t" prefix first and then execute the "P" process later. Sequentialization of prefixes can be used for such purposes.

Sequentialization of prefixes, usually written as the $t \bullet P$ process, indicates that it will perform the "t" prefix first and continue as the "P" process.

3-2 Summation of Processes

The binary operator "$+$", summation, combines two process expressions as alternatives.

For example, the process $P+Q$ can proceed non-deterministically either as the process P or the process Q; as soon as one performs its first action/interaction the other is discarded.

3-3 Parallel Composition of Processes

Parallel composition of two processes P and Q, usually written $P\|Q$, is the key primitive distinguishing the process algebras from sequential models of process executions.

Parallel composition allows the executions in P and Q to proceed simultaneously and independently.

3-4 Recursive Definition of a Process

The operators presented so far describe only finite action/interaction and are consequently insufficient for full computability, which includes non-terminating behavior. Recursion is the operator that allows finite descriptions of infinite behavior.

For example, **fix**($X=E$) can be understood as abbreviating the recursive definition of an infinite behavior denoted by the "X" process variable.

3-5 Replication of a Process

Replication is the other operator that allows finite descriptions of infinite behavior of a process.

For example, replication !P can be understood as abbreviating the parallel composition of a countably infinite number of P processes.

3-6 Conditional Definition of a Process

A process can be defined by a one-or-more-armed conditional expression. For example, the process (**if** $cond_1$ **then** P_1)+(**if** $cond_2$ **then** P_2)...+(**if** $cond_j$ **then** P_j) will proceed as the process P_1 if the "$cond_1$" value is true, or proceed as the process P_2 if the "$cond_2$" value is true,..., or proceed as the process P_j if the "$cond_j$" value is true.

3-7 Null Process

Process algebras generally also include a null process, denoted as *STOP*, which has no interaction points. It is utterly inactive and its sole purpose is to act as the inductive anchor on top of which more interesting processes can be generated.

The process "*STOP*●P_1" (i.e. sequential composition of processes *STOP* and P_1) equals to the process "*STOP*", as shown in Figure 3-1.

$$STOP \bullet P_1 \quad = \quad STOP$$

Figure 3-1   Characteristics of Null Process (I)

The process "$P_2 + STOP$" (i.e. summation of processes P_2 and $STOP$) equals to the process "$STOP + P_2$" (i.e. summation of processes $STOP$ and P_2) which equals to the process "P_2", as shown in Figure 3-2.

$$P_2 + STOP \quad = \quad STOP + P_2 \; = \quad P_2$$

Figure 3-2   Characteristics of Null Process (II)

The process "$P_3 \| STOP$" (i.e. parallel composition of processes P_3 and $STOP$) equals to the process "$STOP \| P_3$" (i.e. parallel composition of processes $STOP$ and P_3) which equals to the process "P_3", as shown in Figure 3-3.

$$P_3 \| STOP \quad = \quad STOP \| P_3 \quad = \quad P_3$$

Figure 3-3   Characteristics of Null Process (III)

PART II: OPERATION-BASED INFINITE-QUEUE SBC PROCESS ALGEBRA

Chapter 4: Operation-Based Value-Passing Interactions

In this chapter, we first introduce operations and operation-based interactions. We then introduce the formal description of an operation-based communication port, an operation-based action and an operation-based interaction.

4-1 Operations

An operation provided by each component represents a procedure, or method, or function of the component as shown in Figure 4-1.

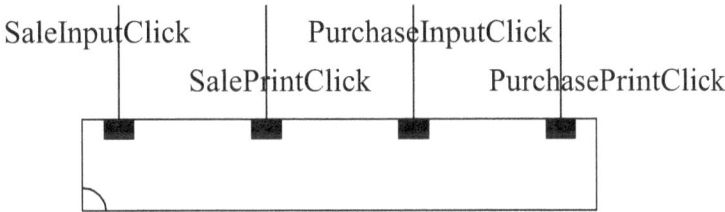

Figure 4-1   An Operation Represents a Procedure, or Method,
or Function of a Component

An operation may contain several input parameters (e.g. i_1, i_2) and output parameters (e.g. o_1, o_2), as shown in Figure 4-2.

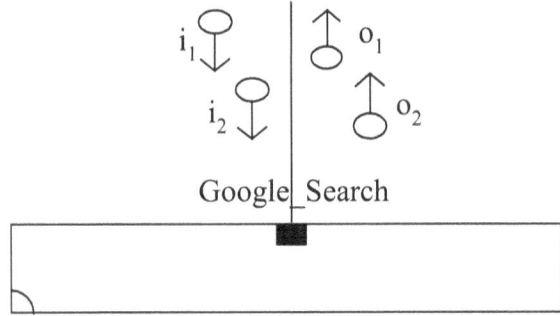

Figure 4-2   An Operation Contains Several Input/Output Parameters

An operation formula is used to completely describe an operation. An operation formula includes a) operation name, b) input parameters (e.g. i_1, i_2, ..., i_m) and c) output parameters (e.g. o_1, o_2, ..., o_n), as shown in Figure 4-3.

$$\text{Operation\_Name (In } i_1, i_2, ..., i_m \text{ ; Out } o_1, o_2, ..., o_n \text{)}$$

Figure 4-3   Operation Formula

4-2 Operation-Based Interactions

An interaction represents an indivisible and instantaneous handshake or communication between two agents. In the operation-based approach as shown in Figure 4-4, the caller agent (either external environment's actor or component) communicates with the callee agent (component) through the operation call or return interaction (also named as operation call or reply message).

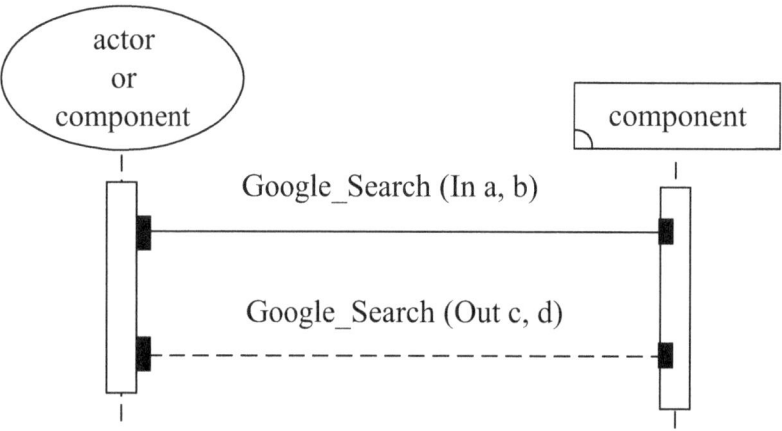

Figure 4-4   Operation-Based Interactions

The caller agent owns the "calling port" of the interaction. In the operation call interaction (also known as operation call message) case, the calling port is "

Google_Search (In a, b)

" and its conduct is to assist the caller agent to output a value to each of the "a" and "b" variables (of the "Google_Search" operation), as shown in Figure 4-5.

Figure 4-5   Calling Port
in the Operation Call Interaction Case

The caller agent together with the "calling port" is named the "calling action" as shown in Figure 4-6.

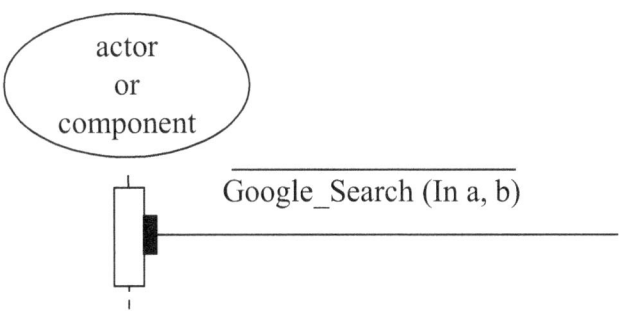

Figure 4-6   Calling Action
in the Operation Call Interaction Case

In the operation return interaction (also known as operation reply message) case, the calling port is " Google_Search (Out c, d) " and its conduct is to assist the caller agent to input a value from each of the "c" and "d" variables (of the "Google_Search" operation), as shown in Figure 4-7.

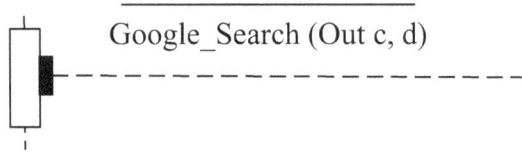

Figure 4-7   Calling Port
in the Operation Return Interaction Case

The caller agent together with the "calling port" is named the "calling action" as shown in Figure 4-8.

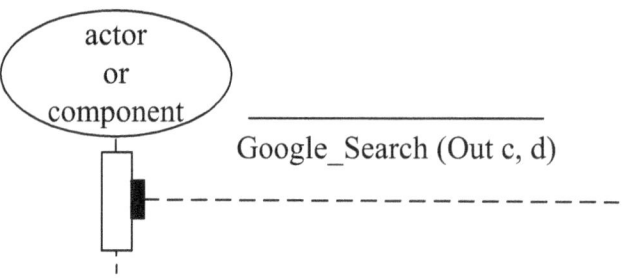

Figure 4-8   Calling Action
in the Operation Return Interaction Case

The callee agent owns the "called port" of the interaction. In the operation call interaction case, the called port is "Google_Search (In a, b)" and its conduct is to assist the callee agent to input a value from each of the "a" and "b" variables (of the "Google_Search" operation), as shown in Figure 4-9.

Figure 4-9   Called Port
in the Operation Call Interaction Case

The callee agent together with the "called port" is named the "called action" as shown in Figure 4-10.

Figure 4-10   Called Action
in the Operation Call Interaction Case

In the operation return interaction case, the called port is "Google_Search (Out c, d)" and its conduct is to assist the callee agent in outputting the values of its "c" and "d" variables (via the "Google_Search" operation), as shown in Figure 4-11.

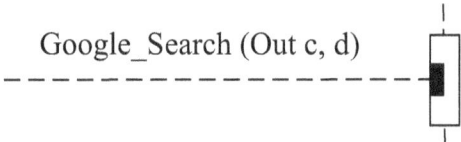

Figure 4-11   Called Port
in the Operation Return Interaction Case

The callee agent together with the "called port" is named the "called action" as shown in Figure 4-12.

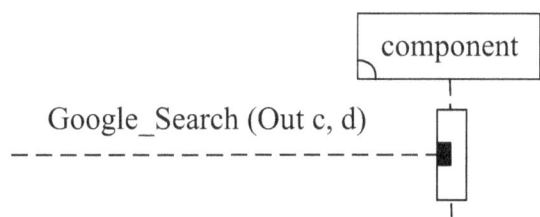

Figure 4-12   Called Action
in the Operation Return Interaction Case

In order to simplify the operation-based interaction diagram, we will redraw it as shown in Figure 4-13.

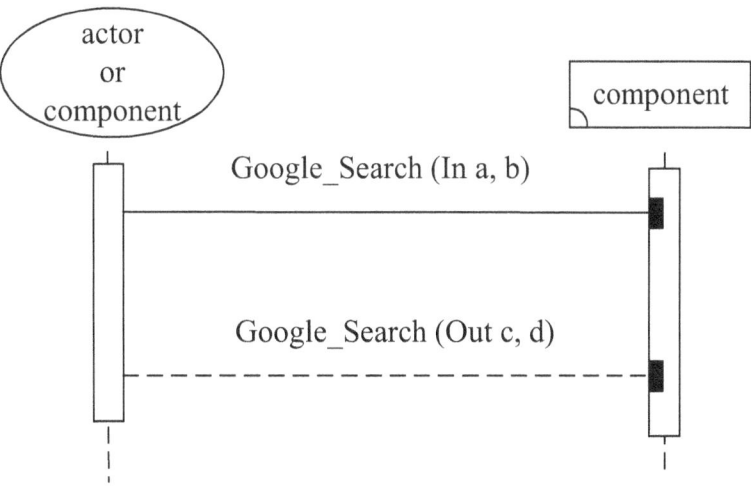

Figure 4-13   Operation-Based Interaction Diagram (I)

Or we can draw the operation-based interaction diagram as shown in Figure 4-14.

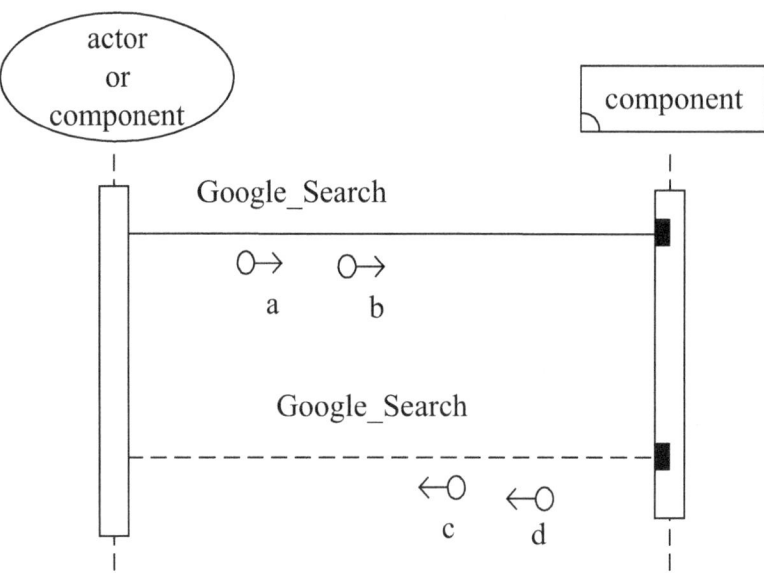

Figure 4-14   Operation-Based Interaction Diagram (II)

We use an operation-based internal interaction (i.e. λ) to represent their handshake or communication, if the caller agent and the callee agent are the same one, as shown in Figure 4-15.

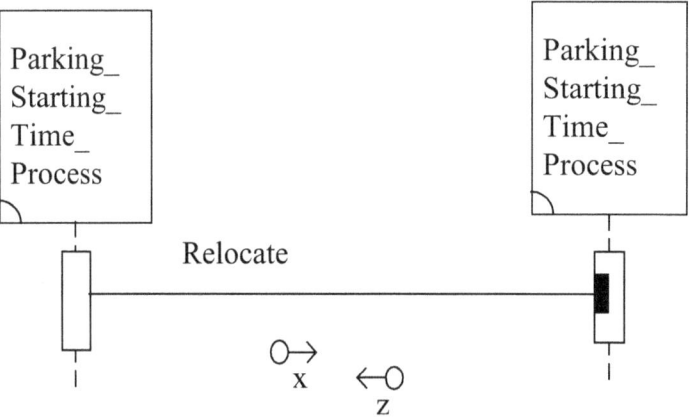

Figure 4-15   An Internal Interaction (I)

Also, we may redraw the internal interaction as shown in Figure 4-16.

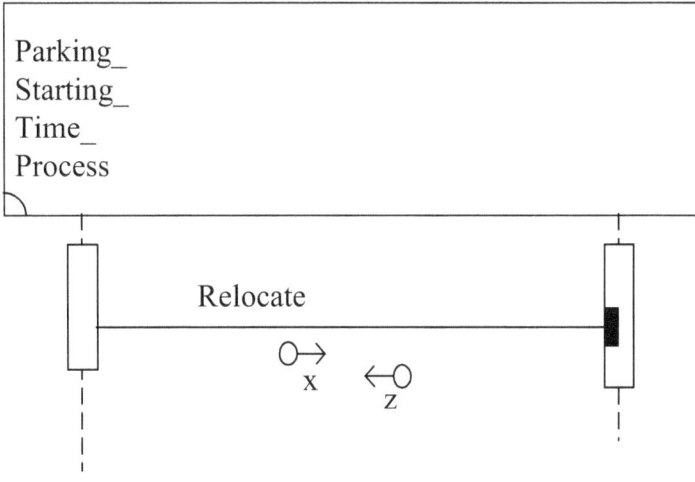

Figure 4-16   An Internal Interaction (II)

4-3 Formal Description of an Operation-Based Communication Port

We formally describe an operation-based communication port as a 3-tuple PORT = <operation_call_or_return, calling_or_called, operation_call_or_return_formula>, where "operation_call_or_return" stands for an OPERATION_CALL or OPERATION_RETURN tag, "calling_or_called" stands for a CALLING or CALLED action tag, and "operation_call_or_return_formula" stands for an operation call or operation return formula as shown in Figure 4-17.

Figure 4-17　Formal Description of
an Operation-Based Communication Port

4-4 Formal Description of an Operation-Based Action

We formally describe an operation-based action as a 4-tuple ACTION = <operation_call_or_return, agent, calling_or_called, operation_call_or_return_formula>, where "operation_call_or_return" stands for an OPERATION_CALL or OPERATION_RETURN tag, "agent" stands for the name of a caller or callee agent, "calling_or_called" stands for a CALLING or CALLED action tag, and "operation_call_or_return_formula" stands for an operation call or operation return formula as shown in Figure 4-18.

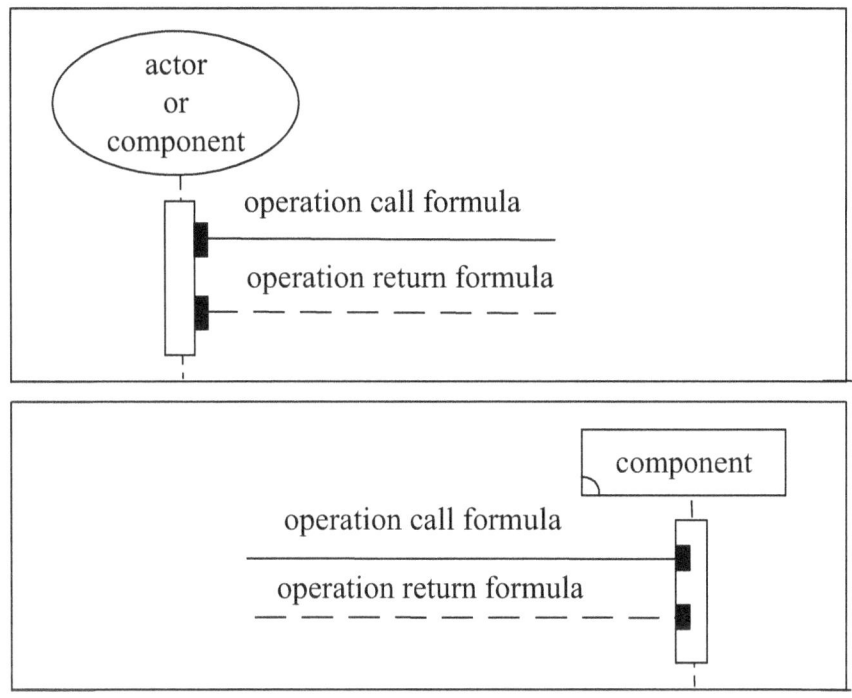

Figure 4-18   Formal Description of an Operation-Based Action

4-5 Formal Description of an Operation-Based Interaction

We formally describe an operation-based interaction as a 4-tuple INTERACTION = <operation_call_or_return, caller_agent, operation_call_or_return_formula, callee_agent>, where "operation_call_or_return" stands for an OPERATION_CALL or OPERATION_RETURN tag, "caller_agent" stands for the name of a caller agent, "operation_call_or_return_formula" stands for an operation call or operation return formula and "callee_agent" stands for the name of a callee agent as shown in Figure 4-19.

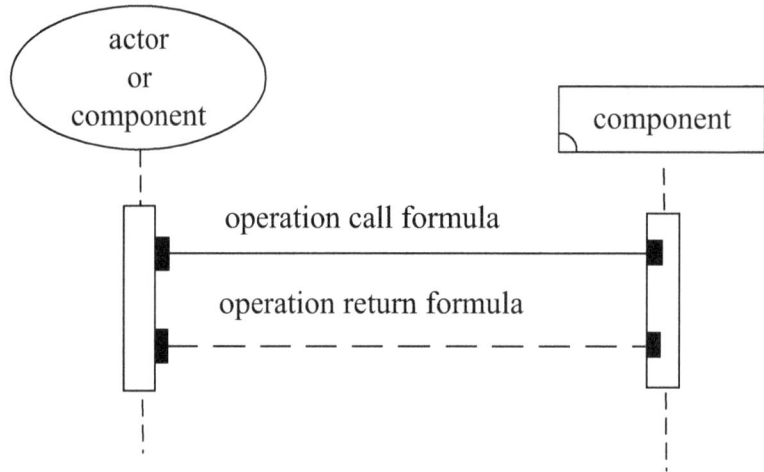

Figure 4-19   Formal Description of an Operation-Based Interaction

Chapter 5: The Structure-Behavior Coalescence Approach

Systems structure and systems behavior are the two most prominent views of a system, integrating the systems structure and systems behavior is apparently the best way to achieve an integrated whole of a system. If we are not able to integrate the systems structure and systems behavior, then there is no way that we are able to integrate the whole system. Structure-behavior coalescence (SBC) provides an elegant way to integrate the systems structure and systems behavior of a system. In other words, SBC facilitates an integrated whole of a system.

5-1 Structure-Behavior Coalescence Means to Integrate the Systems Structure and Systems Behavior

Systems structure, specified by components, their operations and their composition, refers to the type of connection between the components of a system as shown in Figure 5-1.

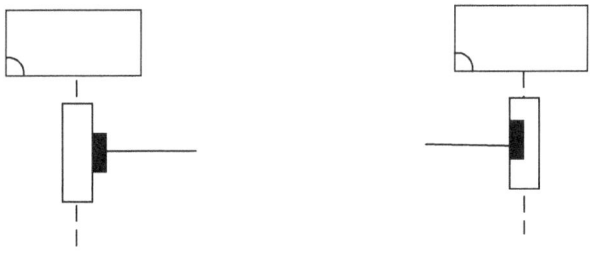

Figure 5-1　Systems Structure

Systems behavior, specified by the interactions between and among the components and environment, refers to the interconnectivities a system in conjunction with its environment as shown in Figure 5-2.

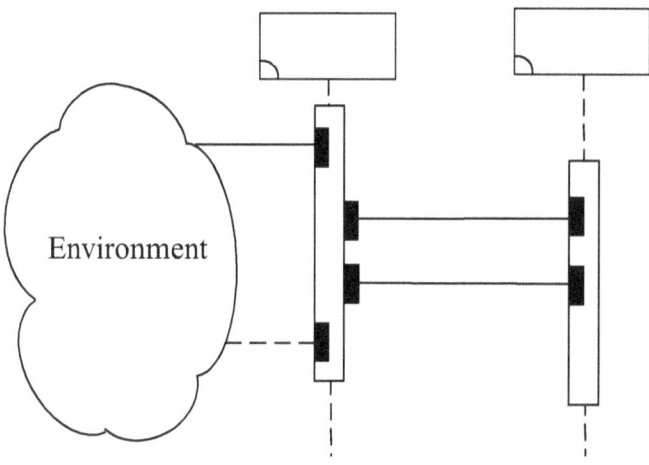

Figure 5-2   Systems Behavior

Systems structure and systems behavior are the two most prominent views of a system, integrating the systems structure and systems behavior is apparently the best way to achieve an integrated whole of a system.

If we are not able to integrate the systems structure and systems behavior, then there is no way that we are able to integrate the whole system.

Structure-behavior coalescence (SBC) [Chao15a] provides an elegant way to integrate the systems structure and systems behavior of a system. In other words, SBC facilitates an integrated whole of a system.

5-2 Interactions among Components and the External Environment to Draw Forth the Systems Behavior

All things that strike us as something independent are essentially parts of a system. We usually call the parts of a system its components. Components are sometimes labeled as parts, entities, objects, building blocks and non-aggregated systems [Chao14a, Chao14b, Chao14c, Chao16a].

In a system, if the components, and among them and the external environment to interact (or handshake), such interaction will draw forth the systems behavior.

A component uses an "action" to interact with the external environment as shown in Figure 5-3.

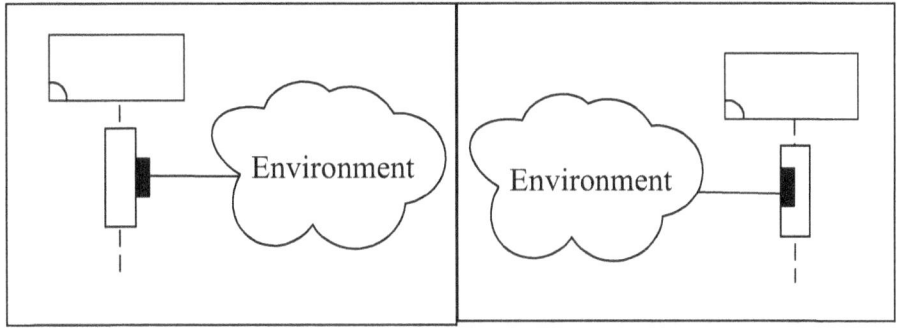

Figure 5-3   A Component Uses an "Action" to Interact
with the External Environment

Two components use an "interaction" to interact with each other as shown in
Figure 5-4.

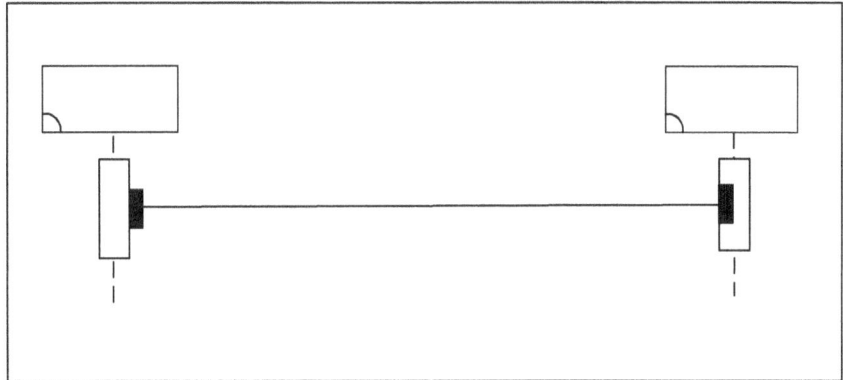

Figure 5-4   Two Components Use an "Interaction"
to Interact with Each Other

5-3 Core Theme of Structure-Behavior Coalescence

The core theme of structure-behavior coalescence is: "Systems Architecture =
Systems Structure -->> Systems Behavior." That is, the systems structure will lead to
the systems behavior as shown in Figure 5-5.

44

Systems Structure X – –>> Systems Behavior X

Figure 5-5   Core Theme of Structure-Behavior Coalescence

One systems structure will draw forth one systems behavior. That is, the systems behavior is attached to or built on the systems structure in the SBC approach.

In other words, the systems behavior can not exist alone; it must be loaded on the systems structure just like a cargo is loaded on a ship as shown in Figure 5-6.

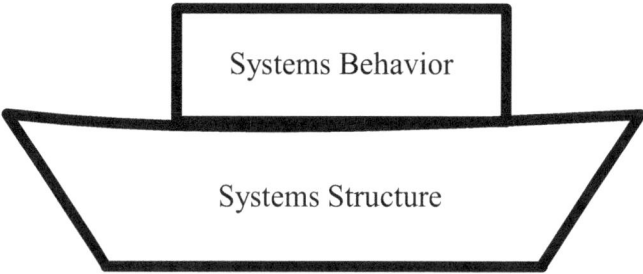

Figure 5-6   Systems Behavior Must be Loaded on the Systems Structure

Chapter 6: Language Constructs of Operation-Based Infinite-Queue SBC Process Algebra

In the chapter, we illustrate in detail those operation-based infinite-queue SBC process algebra language constructs which make up the system modeling of a system.

6-1 Entity Set and Entity Name

As shown in Figure 6-1, we assume an infinite set L of operation call or operation return formulas, and use l_1, l_2...to range over L. Further, we let U be the set of called actions, and use u_1, u_2...to range over U. We let Δ be the set of non-internal interactions, and use a_1, a_2...to range over Δ. We let O be the set of called actions or non-internal interactions, and use o_1, o_2...to range over O. We let T be the set of [condition](called actions/non-internal interactions)[variable settings], and use t_1, t_2 to range over T. Further, we let X be the set of process variables, and use X_1, X_2...to range over X. We let Φ be the set of process Constants, and use A_1, A_2...to range over Φ. We let Π be the set of processes, and use P_1, Q_1...to range over Π. We let Ψ be the set of process expressions, and use E_1, E_2...to range over Ψ. Finally, we let Γ be the set of components, and use C_1, C_2...to range over Γ.

Entity set	Entity name	Type of entity
L	$l_1, l_2...$	operation call or operation return formulas
U	$u_1, u_2...$	called actions
	λ	internal interaction
Δ	$a_1, a_2...$	non-internal interactions
O	$o_1, o_2...$	called actions or non-internal interactions (called actions/non-internal interactions)
T	$t_1, t_2...$	[condition](called actions/non-internal interactions) [variable settings]
X	$X_1, X_2...$	process variables
Φ	$A_1, A_2...$	process Constants
	I, J,...	indexing sets
Π	$P_1, Q_1...$	processes
Ψ	$E_1, E_2...$	process expressions
Γ	$C_1, C_2...$	components

Figure 6-1   Entities

6-2 Backus-Naur Form of Operation-Based Infinite-Queue SBC Processes

The set of operation-based infinite-queue SBC (i.e. Structure-Behavior Coalescence) processes is defined by the following BNF grammar, as shown in Figure 6-2.

(1) E ::= "**!**"<IFD> {"‖ **!** " <IFD>}

(2) <IFD> ::= <Type_1_Expression>
    {"●" <Type_1_Or_2_Expression>} " ● " *STOP*

(3) <Type_1_Or_2_Expression> ::= <Type_1_Expression>

    | <Type_2_Expression>

(4) <Type_1_Expression> ::= <Called-Action>[variable settings]

   $$\Big| \quad \sum_{j \in I} \text{ if } cond_j \text{ then } <\text{Called-Action}_j>[\text{variable settings}_j]$$

(5) <Type_2_Expression> ::= <Interaction>[variable settings]

   $$\Big| \quad \sum_{j \in I} \text{ if } cond_j \text{ then } <\text{Interaction}_j>[\text{variable settings}_j]$$

Figure 6-2   Backus-Naur Form of
Operation-Based Infinite-Queue SBC Processes

6-2-1 Parallel Composition of One or More Replicated Interaction Flow Diagrams Defines the Operation-Based Infinite-Queue SBC Process of a System

Rule 1 describes that parallel composition of one or more replicated (i.e. **!**) interaction flow diagrams (i.e. IFD) defines the operation-based infinite-queue SBC process of a system, as shown in Figure 6-3.

Rule 1
E ::= "**!** "<IFD> {"‖ **!** " <IFD>}

Figure 6-3   Rule 1

6-2-2 An Interaction Flow Diagram is a Type_1 Expression, Followed by Zero or More Type_1_Or_Type_2 Expressions

Rule 2 describes that an interaction flow diagram (i.e. IFD) is defined by a type_1 expression (i.e. Type_1_Expression), followed by zero or more type_1_or_type_2 expressions (i.e. Type_1_Or_2_Expression), as shown in Figure 6-4.

Rule 2
<IFD> ::= <Type_1_Expression> {"● " <Type_1_Or_2_Expression>} " ● " *STOP*

Figure 6-4   Rule 2

6-2-3 Type_1_Or_2 Expression is either Type_1 or Type_2

Rule 3 describes that the type_1_or_2 expression (i.e. Type_1_Or_2_Expression) is either a type_1 expression (i.e. Type_1_Expression) or a type_2 expression (i.e. Type_2_Expression), as shown in Figure 6-5.

Rule 3
<Type_1_Or_2_Expression> ::= <Type_1_Expression> | <Type_2_Expression>

Figure 6-5   Rule 3

6-2-4 Type_1 Expression is either a Called Action with Optional Variable Settings or a Conditional Expression of Called Action with Optional Variable Settings

Rule 4 describes that the type_1 expression (i.e. Type_1_Expression) is either a called action with optional variable settings (i.e. Called-Action[variable settings]) or a conditional expression of called action with optional variable settings (i.e. one-or-more-armed conditional expression of Called-Action$_j$[variable settings$_j$]), as shown in Figure 6-6.

Rule 4
<Type_1_Expression> ::= <Called-Action>[variable settings] $$\left| \sum_{j \in I} \textbf{if } cond_j \textbf{ then } <Called\text{-}Action_j>[variable\ settings_j]\right.$$

Figure 6-6   Rule 4

6-2-5 Type_2 Expression is either an Interaction with Optional Variable Settings or a Conditional Expression of Interaction with Optional Variable Settings

Rule 5 describes that the type_2 expression (i.e. Type_2_Expression) is either an interaction with optional variable settings (i.e. Interaction[variable settings]) or a conditional expression of interaction with optional variable settings (i.e. one-or-more-armed conditional expression of Interaction$_j$[variable settings$_j$]), as shown in Figure 6-7.

Rule 5
<Type_2_Expression> ::= <Interaction>[variable settings] $$\left| \sum_{j \in I} \textbf{if } cond_j \textbf{ then } <Interaction_j>[variable\ settings_j]\right.$$

Figure 6-7   Rule 5

Chapter 7: Transitional Semantics of Operation-Based Infinite-Queue SBC Process Algebra

In the chapter, we illustrate in detail those operation-based infinite-queue SBC process algebra transitional semantics which regards the system modeling of a system.

7-1 Transitional Semantics

In giving meaning to the operation-based infinite-queue SBC process algebra, we shall use the following labelled transition system (LTS)

$$(\Psi, T, \rightarrow)$$

which consists of a set Ψ of process expressions, a set T of "[condition](called actions/non-internal interactions)[variable settings]", and a transition relation $\rightarrow \subseteq \Psi \times T \times \Psi$ where $(E_i, t, E_j) \in \rightarrow$ is denoted by $E_i \xrightarrow{t} E_j$.

The semantics for Ψ consists in the transition rules of each transition relation $\rightarrow$ over $\Psi \times T \times \Psi$. These transition rules will follow the structure of expressions.

As shown in Figure 7-1, we give the complete set of transition rules; the names Prefix, Sum, Parallel, and Constant indicate that the rules are associated respectively with Prefix, Summation, and Parallel Composition and with Constants.

$$\text{Prefix} \qquad \frac{\rule{3cm}{0.4pt}}{t \bullet E \xrightarrow{t} E}$$

$$\text{Sum}_j \qquad \frac{E_j \xrightarrow{t} E'_j}{\sum_{i \in I} E_i \xrightarrow{t} E'_j} (j \in I)$$

$$\text{Parallel}_1 \qquad \frac{E \xrightarrow{t} E'}{E \parallel F \xrightarrow{t} E' \parallel F}$$

$$\text{Parallel}_2 \qquad \frac{F \xrightarrow{t} F'}{E \parallel F \xrightarrow{t} E \parallel F'}$$

$$\text{Constant} \qquad \frac{P \xrightarrow{t} P'}{A \xrightarrow{t} P'} \quad (A \overset{\text{def}}{=} P)$$

Figure 7-1   Transition Rules for the
Operation-Based Infinite-Queue SBC Process Algebra

7-2 Rule of Prefix

The rule for Prefix, shown in Figure 7-2, can be read as follows: Under any circumstances, we always infer $t \bullet E \xrightarrow{t} E$. That is, an expression, with a [condition](called action/non-internal interaction)[variable settings] prefixed to it, will use this [condition](called action/non-internal interaction)[variable settings] to accomplish the transition.

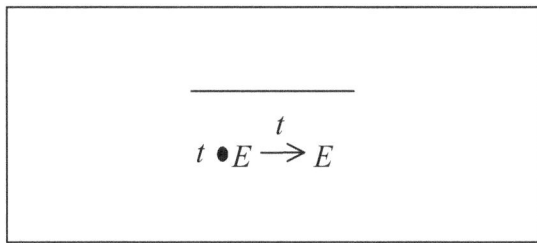

Figure 7-2   Rule of Prefix

7-3 Rule of Summation

The rule for Summation, shown in Figure 7-3, can be read as follows: If any one summand E_j of the sum $\sum_{i \in I} E_i$ has a [condition](called action/non-internal interaction)[variable settings], then the whole sum also has that [condition](called action/non-internal interaction)[variable settings].

$$\frac{E_j \xrightarrow{t} E'_j}{\sum_{i \in I} E_i \xrightarrow{t} E'_j} (j \in I)$$

Figure 7-3   Rule of Summation

7-4 Rule of Parallel Composition

There are two transition rules for parallel composition. Rule Parallel$_1$, as shown in Figure 7-4, indicates that from $E \xrightarrow{t} E'$ we shall infer $E\|F \xrightarrow{t} E'\|F$.

$$\frac{E \xrightarrow{t} E'}{E \parallel F \xrightarrow{t} E' \parallel F}$$

Figure 7-4   Rule Parallel$_1$

Rule Parallel$_2$, as shown in Figure 7-5, indicates that from $F \xrightarrow{t} F'$ we shall infer $E \parallel F \xrightarrow{t} E \parallel F'$.

$$\frac{F \xrightarrow{t} F'}{E \parallel F \xrightarrow{t} E \parallel F'}$$

Figure 7-5   Rule Parallel$_2$

7-5 Rule of Constants

The rule for Constants, shown in Figure 7-6, can be read as follows: the rule of Constants asserts that each Constant has the same transitions as its defining expression.

$$\frac{P \xrightarrow{t} P'}{A \xrightarrow{t} P'} \quad (A \overset{\text{def}}{=} P)$$

Figure 7-6   Rule of Constants

PART III: CASES STUDY

Chapter 8: Operation-Based Infinite-Queue SBC Process of the Robot System

In this chapter, we use the operation-based infinite-queue SBC process algebra to model the *Robot* system as shown in Figure 8-1.

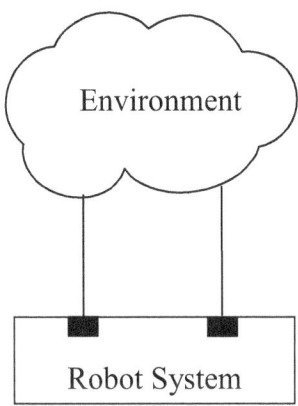

Figure 8-1   Systems Modeling the *Robot* System

8-1 BNF Tree of the Robot System

The operation-based infinite-queue SBC process of the *Robot* system, A_{001}, is defined as "$!(t_{001} \bullet t_{002} \bullet STOP) \| !(t_{003} \bullet t_{004} \bullet STOP)$".

We draw the operation-based infinite-queue SBC process algebra Backus-Naur Form tree of A_{001} as shown in Figure 8-2.

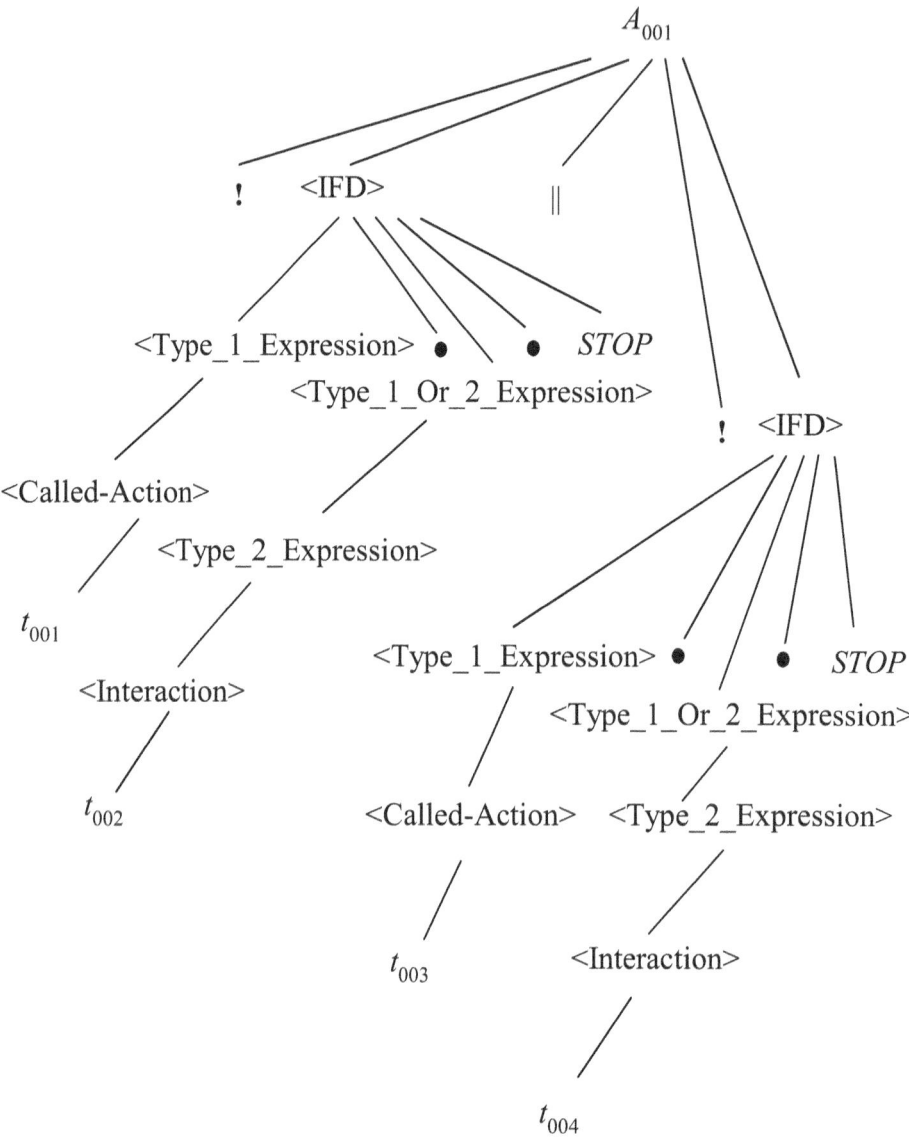

Figure 8-2   Backus-Naur Form Tree of the *Robot* System's
Operation-Based Infinite-Queue SBC Process

There are two IFDs in the operation-based infinite-queue SBC process of the
Robot system. The first IFD is shown in Figure 8-3.

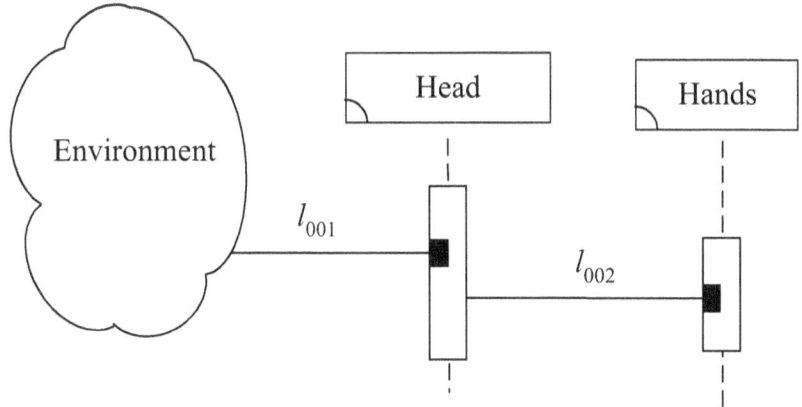

Figure 8-3   First IFD of the *Robot* System

The second IFD of the operation-based infinite-queue SBC process of the *Robot* system is shown in Figure 8-4.

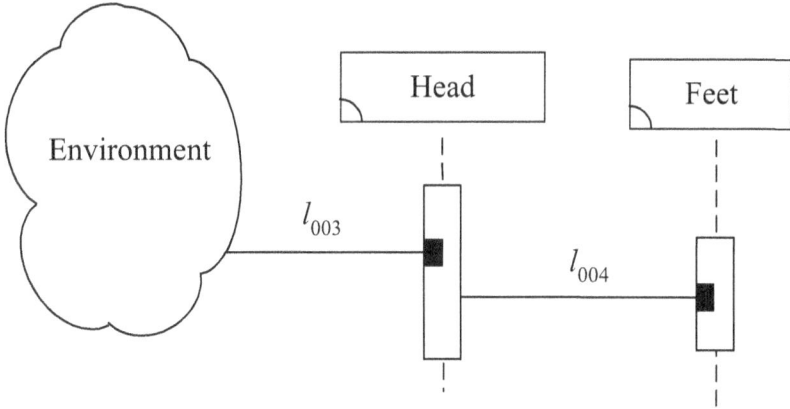

Figure 8-4   Second IFD of the *Robot* System

8-2 Prefixes of the Robot System

The t_{001} (operation-based called action with no conditions and no variable settings) prefix is defined as "<OPERATION_CALL, *Head*, CALLED, l_{001}>", as shown in Figure 8-5.

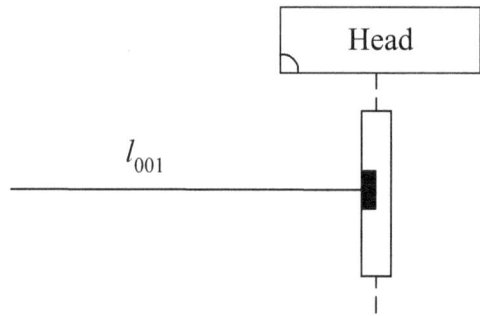

Figure 8-5   Prefix of t_{001}

The t_{002} (operation-based interaction with no conditions and no variable settings) prefix is defined as "<OPERATION_CALL, *Head*, l_{002}, *Hands*>", as shown in Figure 8-6.

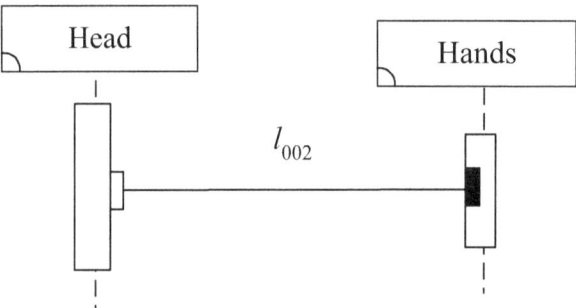

Figure 8-6   Prefix of t_{002}

The t_{003} (operation-based called action with no conditions and no variable settings) prefix is defined as "<OPERATION_CALL, *Head*, CALLED, l_{003}>", as shown in Figure 8-7.

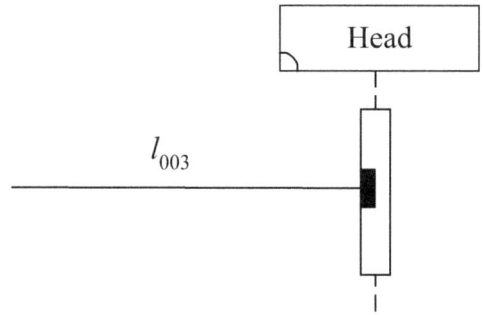

Figure 8-7   Prefix of t_{003}

The t_{004} (operation-based interaction with no conditions and no variable settings) prefix is defined as "<OPERATION_CALL, *Head*, l_{004}, *Feet*>", as shown in Figure 8-8.

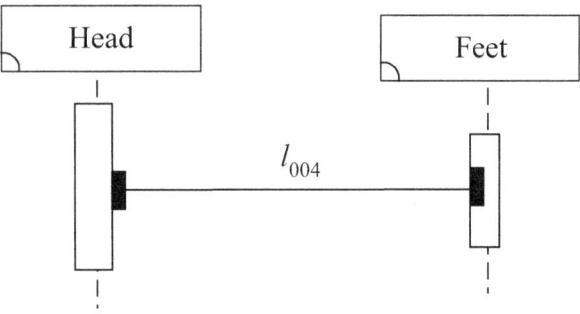

Figure 8-8   Prefix of t_{004}

Figure 8-9 shows all operation call or operation return formulas of the operation-based infinite-queue SBC process of the *Robot* system.

Entity name	Operation Call Formula	Operation Return Formula
l_{001}	Receive_Write_Signal	
l_{002}	Move_Hand	
l_{003}	Receive_Wall_Signal	
l_{004}	Move_Foot	

Figure 8-9   Operation Call or Operation Return Formulas
of the *Robot* System's Process

8-3 Process of the Robot System

The following transition graph shows, in Figure 8-10, the semantics of P_{001}'s operation-based infinite-queue SBC process.

62

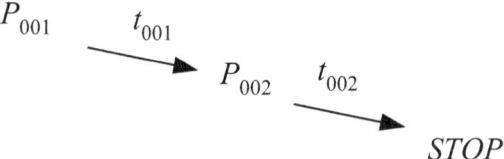

Figure 8-10   Transition graph of the P_{001} Process

In the transition graph of the P_{001}'s operation-based infinite-queue SBC process, processes P_{001} and P_{002} are defined as in Figure 8-11.

$$P_{001} \stackrel{\text{def}}{=\!=} t_{001} \bullet P_{002}$$

$$P_{002} \stackrel{\text{def}}{=\!=} t_{002} \bullet STOP$$

Figure 8-11   Definition of Processes P_{001} and P_{002}

The following transition graph shows, in Figure 8-12, the semantics of P_{003}'s operation-based infinite-queue SBC process.

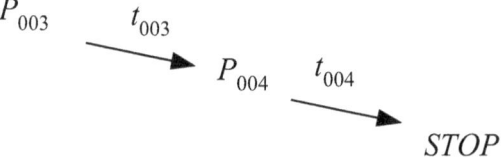

Figure 8-12   Transition graph of the P_{003} Process

In the transition graph of the P_{003}'s operation-based infinite-queue SBC process, processes P_{003} and P_{004} are defined as in Figure 8-13.

$$P_{003} \stackrel{\text{def}}{=} t_{003} \bullet P_{004}$$

$$P_{004} \stackrel{\text{def}}{=} t_{004} \bullet STOP$$

Figure 8-13   Definition of Processes P_{003} and P_{004}

The operation-based infinite-queue SBC process of the *Robot* system, A_{001}, is defined as "$!P_{001}\|!P_{003}$".

Chapter 9: Operation-Based Infinite-Queue SBC Process of the Multi-Tier Personal Data System

After the systems development is finished, the *Multi-Tier Personal Data System* shall appear on a multi-tier platform [Wall04] as shown in Figure 9-1.

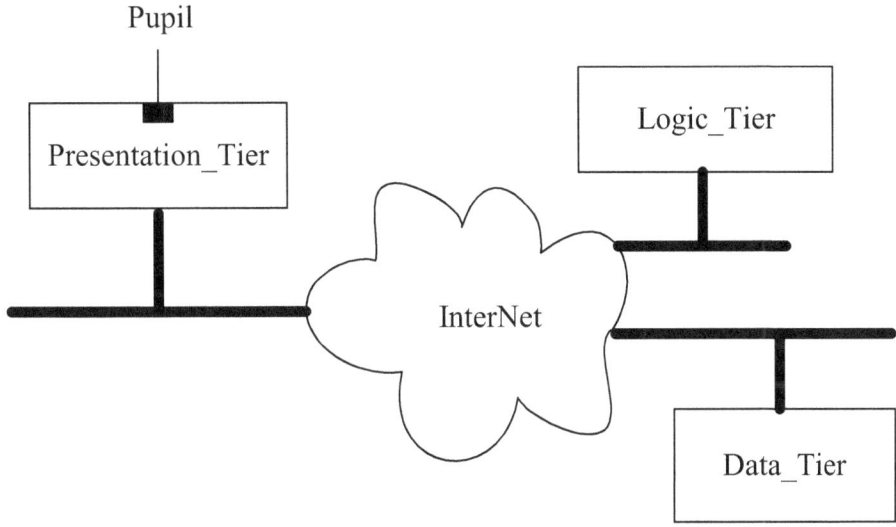

Figure 9-1   *Multi-Tier Personal Data System* on a Multi-Tier Platform

In the *Data_Tier*, there is a *Personal_Database* database [Date03, Elma10] which contains a *Personal_Data* table as shown in Figure 9-2.

Social_Secuirty_Number	Name	Date_of_Birth	Sex	Height (cm)	Weight (Kg)
318-49-2465	Mary R. Williams	June 17, 1976	Female	165	51
424-87-3651	Lee H. Wulf	July 24, 1982	Female	162	76
512-24-3722	John K. Bryant	May 12, 1954	Male	180	80

Record: 4 of 4

Figure 9-2   *Personal_Database* Contains *Personal_Data*

The functionality of the *Multi-Tier Personal Data System* is to provide a graphical user interface (GUI) [Gali07] for the external actor to trigger two behaviors. The first behavior is *AgeCalculation* and the second behavior is *OverweightCalculation*, as shown in Figure 9-3.

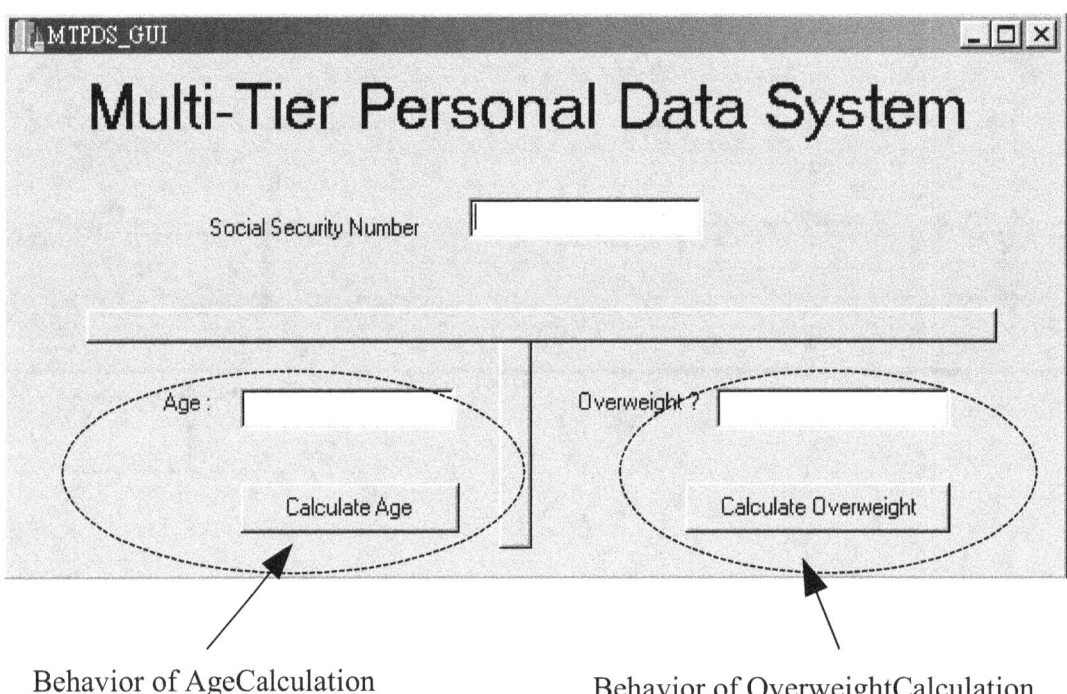

Figure 9-3   Two Behaviors

In the *AgeCalculation* behavior, the external actor inputs an integer *Social_Security_Number* value then presses down the *Calculate_Age* button. After that, the *Multi-Tier Personal Data System* retrieves the *Date_of_Birth* value from the database in line with the corresponding *Social_Security_Number* value. From the *Date_of_Birth* value, the *Multi-Tier Personal Data System* calculates the *Age* value and displays it on the screen. Figure 9-4 shows the *Social_Security_Number* value is 512-24-3722 and the retrieved *Date_of_Birth* value is May 12, 1954 and the calculated *Age* value, which is 59, is then displayed on the screen.

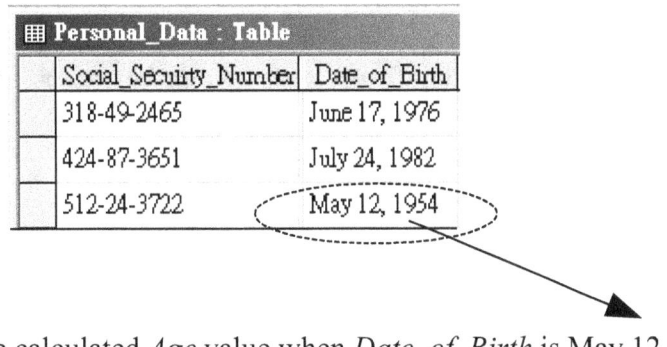

the calculated *Age* value when *Date_of_Birth* is May 12, 1954

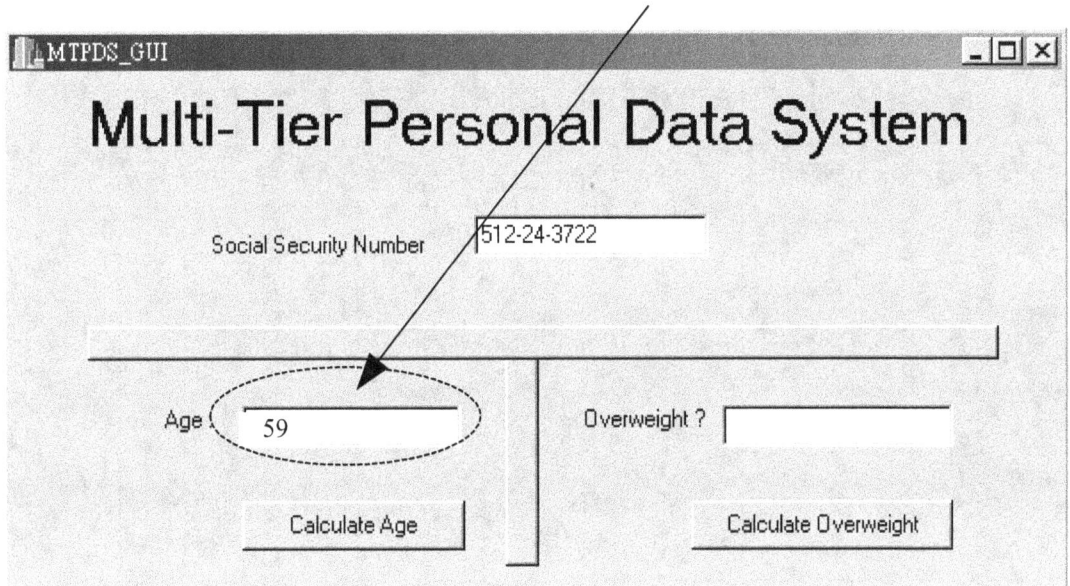

Figure 9-4   Behavior of *AgeCalculation*

In the *OverweightCalculation* behavior, the external actor inputs an integer *Social_Security_Number* value then presses down the *Calculate_Overweight* button. After that, the *Multi-Tier Personal Data System* retrieves the *Weight, Height* and *Sex* values from the database in line with the corresponding *Social_Security_Number* value. From the *Weight, Height* and *Sex* values, the *Multi-Tier Personal Data System* calculates the true-or-false *Overweight* value and displays it on the screen. Figure 9-5 shows the *Social_Security_Number* value is 318-49-2465 and the retrieved Sex, H*eight* and *Weight* values are Female, 165 and 51, respectively, the calculated *Overweight* value, which is "*No*", is then displayed on the screen.

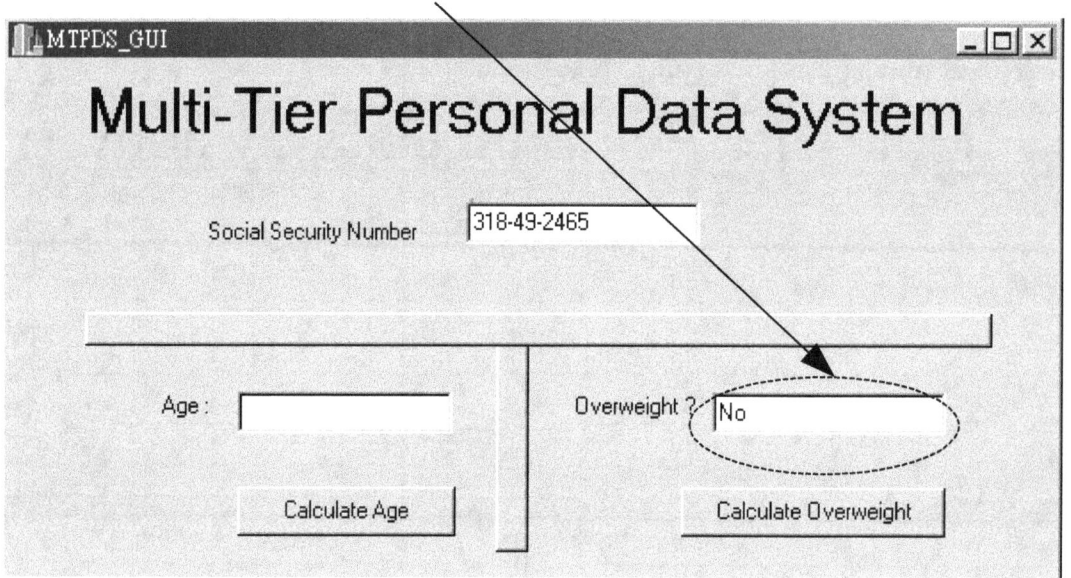

Social_Securty_Number	Sex	Height (cm)	Weight (Kg)
318-49-2465	Female	165	51
424-87-3651	Female	162	76
512-24-3722	Male	180	80

Calculated Overweight when Sex, Height (cm), and Weight (kg) are Female, 165, and 51

Figure 9-5   Behavior of *OverweightCalculation*

In this chapter, we use the operation-based infinite-queue SBC process algebra to model the *Multi-Tier Personal Data System* as shown in Figure 9-6.

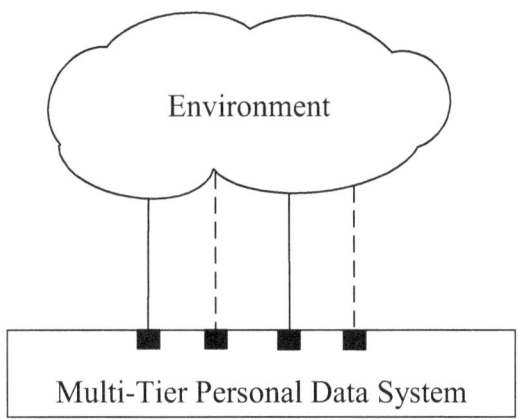

Figure 9-6   Systems Modeling of
the *Multi-Tier Personal Data System*

9-1 BNF Tree of the Multi-Tier Personal Data System

The operation-based infinite-queue SBC process of the *Multi-Tier Personal Data System*, A_{011}, is defined as "$!(t_{011} \bullet t_{012} \bullet t_{013} \bullet t_{014} \bullet t_{015} \bullet STOP) \| !(t_{016} \bullet t_{017} \bullet t_{018} \bullet t_{019} \bullet t_{020} \bullet STOP)$".

We draw the operation-based infinite-queue SBC process algebra Backus-Naur Form tree of A_{011} as shown in Figure 9-7.

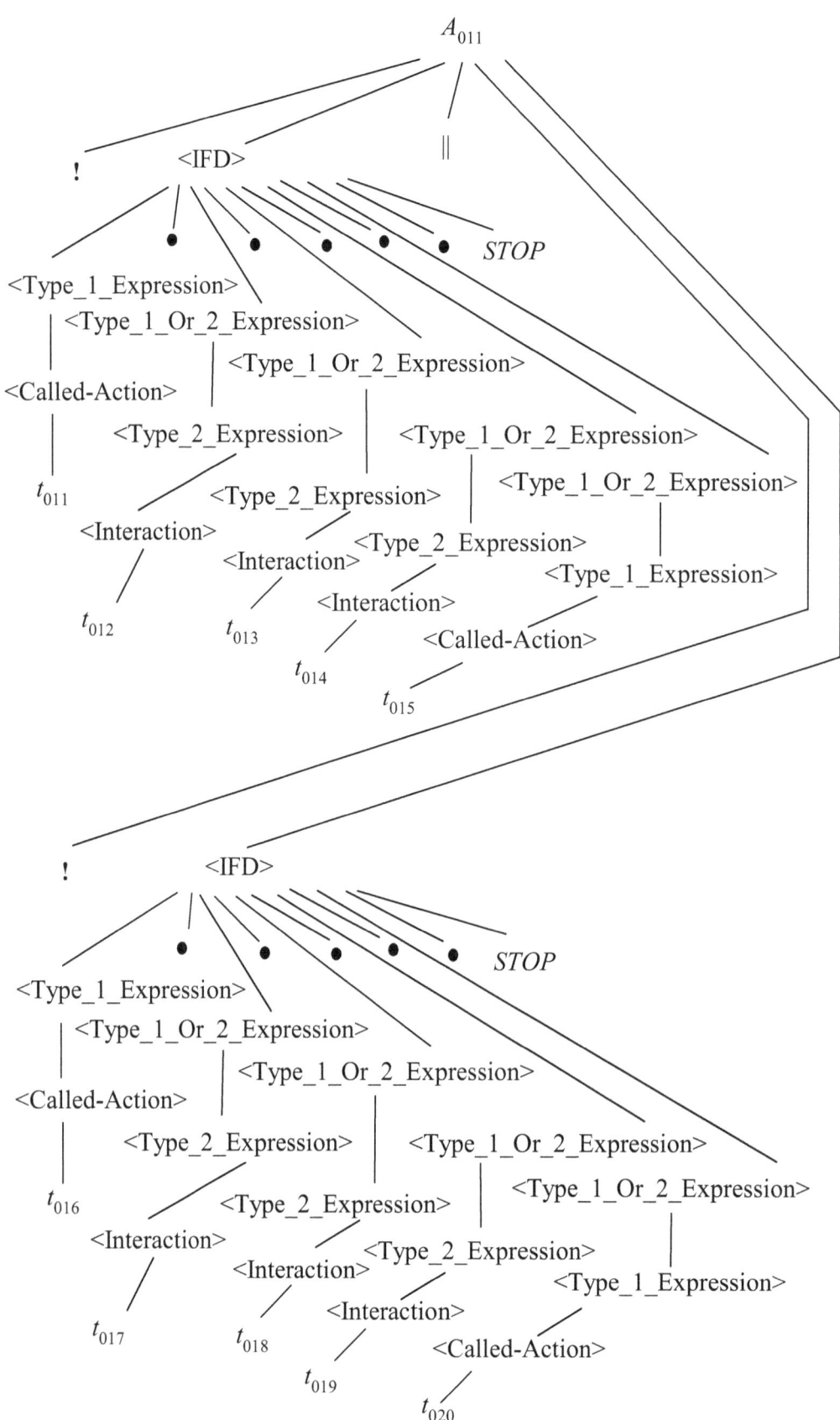

Figure 9-7   Backus-Naur Form Tree of the *Multi-Tier Personal Data System*'s
Operation-Based Infinite-Queue SBC Process

There are two IFDs in the operation-based infinite-queue SBC process of the *Multi-Tier Personal Data System*. The first IFD is shown in Figure 9-8.

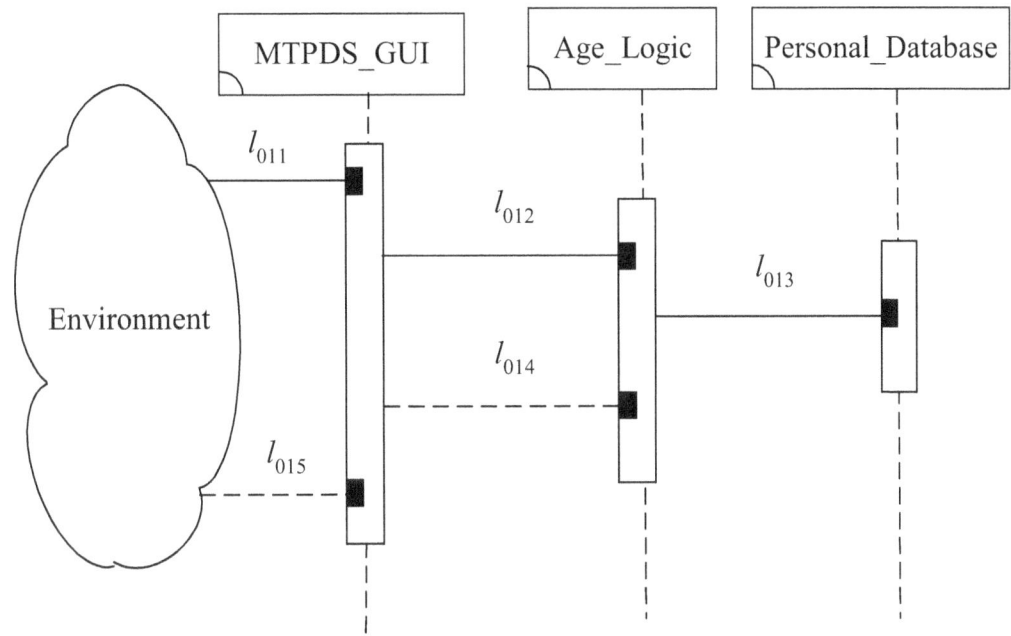

Figure 9-8　First IFD of the *Multi-Tier Personal Data System*

The second IFD of the operation-based infinite-queue SBC process of the *Multi-Tier Personal Data System* is shown in Figure 9-9.

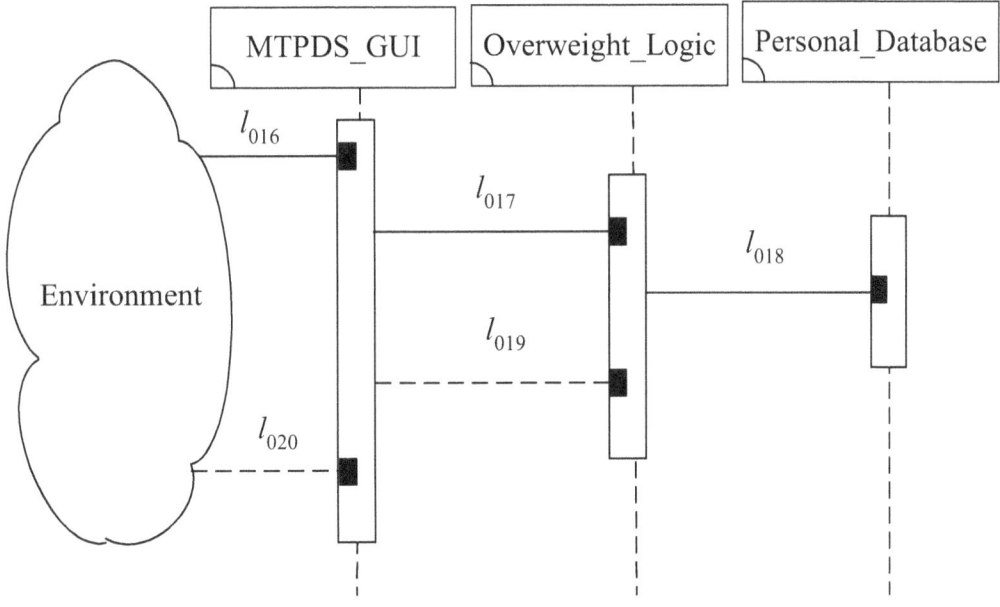

Figure 9-9　Second IFD of the *Multi-Tier Personal Data System*

9-2 Prefixes of the Multi-Tier Personal Data System

The t_{011} (operation-based called action with no conditions and no variable settings) prefix is defined as "<OPERATION_CALL, *MTPDS_GUI*, CALLED, l_{011}>", as shown in Figure 9-10.

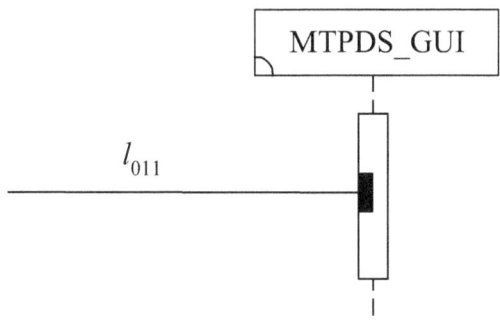

Figure 9-10   Prefix of t_{011}

The t_{012} (operation-based interaction with no conditions and no variable settings) prefix is defined as "<OPERATION_CALL, *MTPDS_GUI*, l_{012}, *Age_Logic*>", as shown in Figure 9-11.

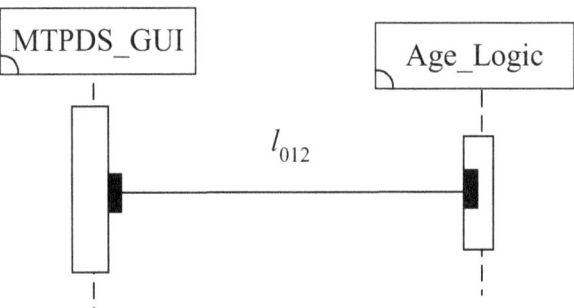

Figure 9-11   Prefix of t_{012}

The t_{013} (operation-based interaction with no conditions and no variable settings) prefix is defined as "<OPERATION_CALL, *Age_Logic*, l_{013}, *Personal_Database*>", as shown in Figure 9-12.

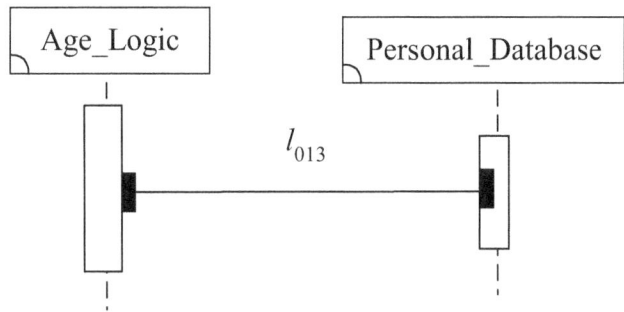

Figure 9-12   Prefix of t_{013}

The t_{014} (operation-based interaction with no conditions and no variable settings) prefix is defined as "<OPERATION_RETURN, *MTPDS_GUI*, l_{014}, *Age_Logic*>", as shown in Figure 9-13.

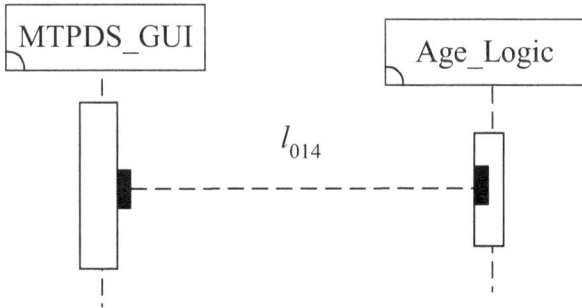

Figure 9-13   Prefix of t_{014}

The t_{015} (operation-based called action with no conditions and no variable settings) prefix is defined as "<OPERATION_RETURN, *MTPDS_GUI*, CALLED, l_{015}>", as shown in Figure 9-14.

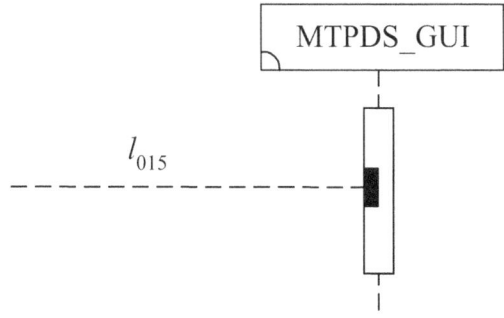

Figure 9-14   Prefix of t_{015}

The t_{016} (operation-based called action with no conditions and no variable settings) prefix is defined as "<OPERATION_CALL, *MTPDS_GUI*, CALLED, l_{016}>", as shown in Figure 9-15.

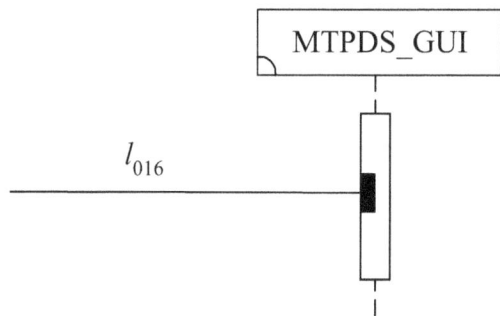

Figure 9-15   Prefix of t_{016}

The t_{017} (operation-based interaction with no conditions and no variable settings) prefix is defined as "<OPERATION_CALL, *MTPDS_GUI*, l_{017}, *Overweight_Logic*>", as shown in Figure 9-16.

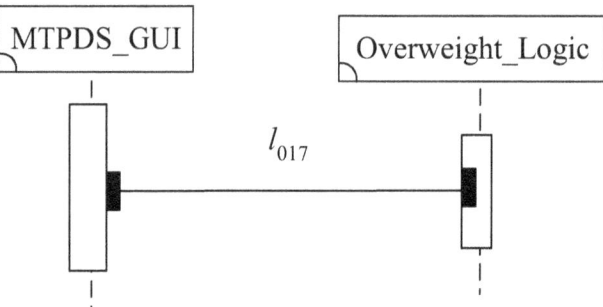

Figure 9-16   Prefix of t_{017}

The t_{018} (operation-based interaction with no conditions and no variable settings) prefix is defined as "<OPERATION_CALL, *Overweight_Logic*, l_{018}, *Personal_Database*>", as shown in Figure 9-17.

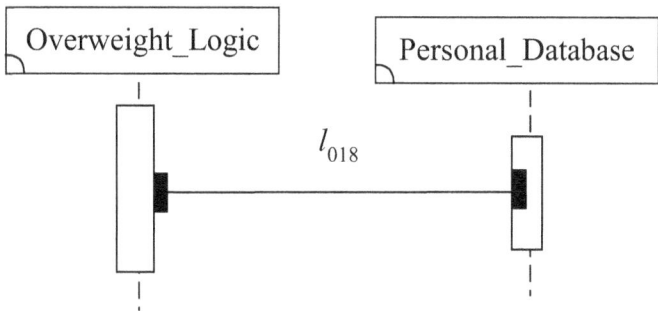

Figure 9-17　Prefix of t_{018}

The t_{019} (operation-based interaction with no conditions and no variable settings) prefix is defined as "<OPERATION_RETURN, *MTPDS_GUI*, l_{019}, *Overweight_Logic*>", as shown in Figure 9-18.

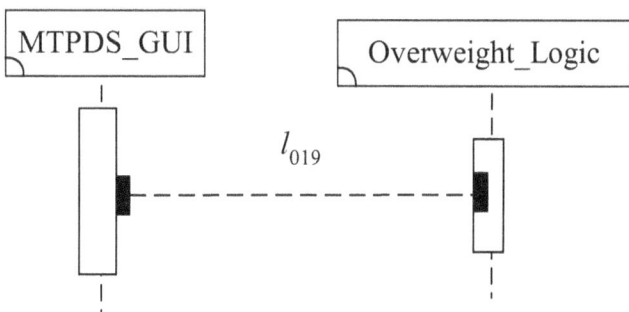

Figure 9-18　Prefix of t_{019}

The t_{020} (operation-based called action with no conditions and no variable settings) prefix is defined as "<OPERATION_RETURN, *MTPDS_GUI*, CALLED, l_{020}>", as shown in Figure 9-19.

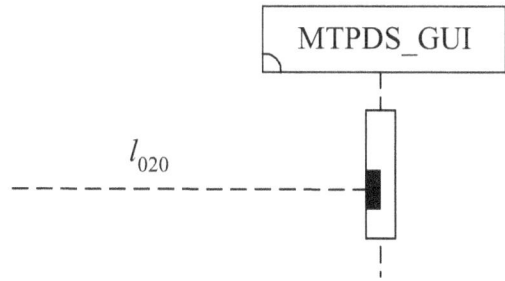

Figure 9-19　Prefix of t_{020}

Figure 9-20 shows all operation call or operation return formulas of the operation-based infinite-queue SBC process of the *Multi-Tier Personal Data System.*

Entity name	Operation Call Formula	Operation Return Formula
l_{011}	Calculate_AgeClick(In Social_Security_Number)	
l_{012}	Calculate_Age(In Social_Security_Number)	
l_{013}	Sql_DateOfBirth_Select(In Social_Security_Number; Out query_DateOfBirth)	
l_{014}		Calculate_Age(Out Age)
l_{015}		Calculate_AgeClick(Out Age)
l_{016}	Calculate_OverweightClick(In Social_Security_Number)	
l_{017}	Calculate_Overweight(In Social_Security_Number)	
l_{018}	Sql_SexHeightWeight_Select(In Social_Security_Number; Out query_SexHeightWeight)	
l_{019}		Calculate_Overweight(Out Overweight)
l_{020}		Calculate_OverweightClick(Out Overweight)

Figure 9-20   Operation Call or Operation Return Formulas
of the *Multi-Tier Personal Data System*'s Process

Figure 9-21 shows the primitive data type specification of the *Social_Security_Number* input parameter and the *Age, Overweight* output parameters.

Parameter	Data Type	Instances
Social_Security_Number	Text	424-87-3651, 512-24-3722
Age	Integer	28, 56
Overweight	Boolean	Yes, No

Figure 9-21   Primitive Data Type Specification

Figure 9-22 shows the composite data type specification of the *query_DateOfBirth* output parameter occurring in the *Sql_DateOfBirth_Select(In Social_Security_Number; Out query_DateOfBirth)* operation call formula.

Parameter	*query_DateOfBirth*

Data Type	TABLE of
	   Social_Security_Number : Text
	   Age : Integer
	End TABLE;

Instances	424-87-3651 28 512-24-3722 56

Figure 9-22   Composite Data Type Specification

Figure 9-23 shows the composite data type specification of the *query_SexHeightWeight* output parameter occurring in the *Sql_SexHeightWeight_Select(In Social_Security_Number; Out query_SexHeightWeight)* operation call formula.

Parameter	*query_SexHeightWeight*

Data Type	TABLE of
	   Social_Security_Number : Text
	   Sex : Text
	   Height : Number
	   Weight : Number
	End TABLE;

Instances	<table><tr><td>424-87-3651</td><td>Female</td><td>162</td><td>76</td></tr></table> <table><tr><td>512-24-3722</td><td>Male</td><td>180</td><td>80</td></tr></table>

Figure 9-23   Composite Data Type Specification

9-3 Process of the Multi-Tier Personal Data System

The following transition graph shows, in Figure 9-24, the semantics of P_{011}'s operation-based infinite-queue SBC process.

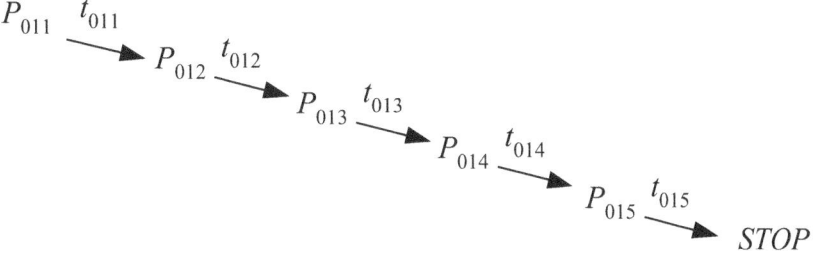

Figure 9-24   Transition graph of the P_{011} Process

In the transition graph of the P_{011}'s operation-based infinite-queue SBC process, processes P_{011}, P_{012}, P_{013}, P_{014} and P_{015} are defined as in Figure 9-25.

$$P_{011} \overset{\text{def}}{=\!=} t_{011} \bullet P_{012}$$

$$P_{012} \overset{\text{def}}{=\!=} t_{012} \bullet P_{013}$$

$$P_{013} \overset{\text{def}}{=\!=} t_{013} \bullet P_{014}$$

$$P_{014} \overset{\text{def}}{=\!=} t_{014} \bullet P_{015}$$

$$P_{015} \overset{\text{def}}{=\!=} t_{015} \bullet STOP$$

Figure 9-25   Definition of Processes
$P_{011}, P_{012}, P_{013}, P_{014}$, and P_{015}

The following transition graph shows, in Figure 9-26, the semantics of P_{016}'s operation-based infinite-queue SBC process.

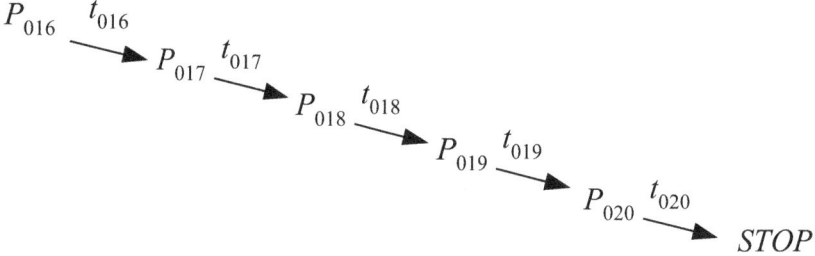

Figure 9-26   Transition graph of the P_{016} Process

In the transition graph of the P_{016}'s operation-based infinite-queue SBC process, processes P_{016}, P_{017}, P_{018}, P_{019} and P_{020} are defined as in Figure 9-27.

$$P_{016} \overset{\text{def}}{=\!=} t_{016} \bullet P_{017}$$

$$P_{017} \overset{\text{def}}{=\!=} t_{017} \bullet P_{018}$$

$$P_{018} \overset{\text{def}}{=\!=} t_{018} \bullet P_{019}$$

$$P_{019} \overset{\text{def}}{=\!=} t_{019} \bullet P_{020}$$

$$P_{020} \overset{\text{def}}{=\!=} t_{020} \bullet STOP$$

Figure 9-27   Definition of Processes
P_{016}, P_{017}, P_{018}, P_{019}, and P_{020}

The operation-based infinite-queue SBC process of the *Multi-Tier Personal Data System*, A_{011}, is defined as "$!P_{011} \| !P_{016}$".

Chapter 10: Operation-Based Infinite-Queue SBC Process of the Mathematical Calculation System

The functionality of the *Mathematical Calculation System* is to provide a graphical user interface (GUI) [Gali07] for the external actor to trigger two behaviors. The first behavior is *DIVIDE&MAXIMUM* and the second behavior is *DIVIDE&GCD*, as shown in Figure 10-1.

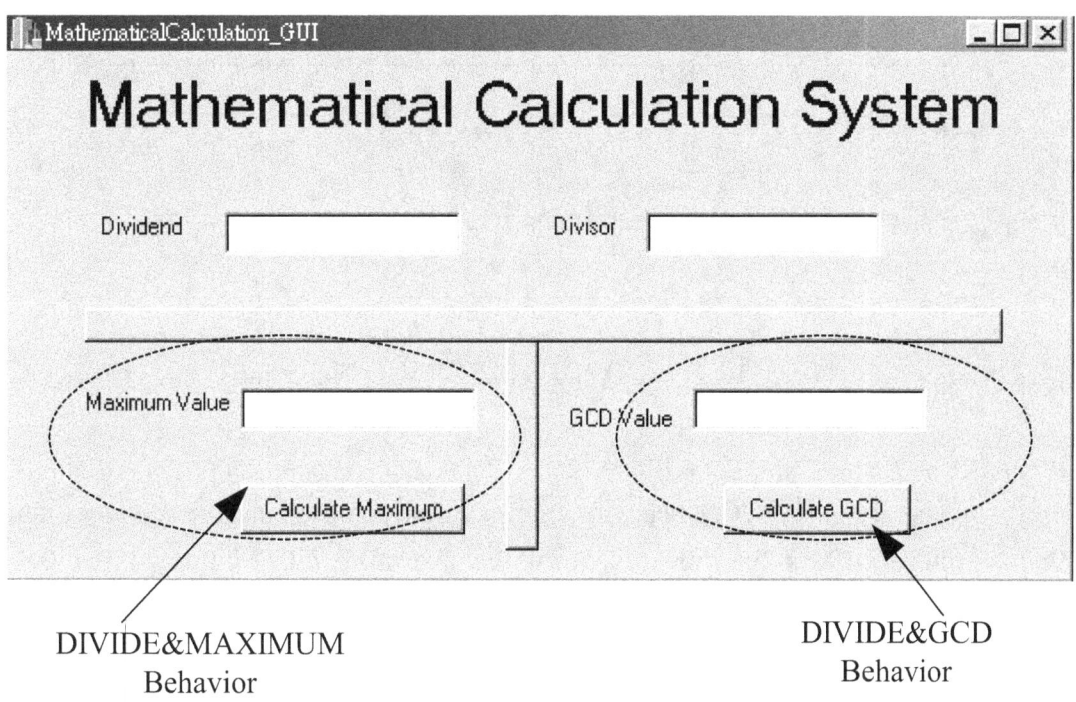

DIVIDE&MAXIMUM
Behavior

DIVIDE&GCD
Behavior

Figure 10-1   Two Behaviors

In the *DIVIDE&MAXIMUM* behavior, the external actor inputs the *Dividend* and *Divisor* integer values then presses down the *Calculate_Maximum* button. After that, *Mathematical Calculation System* will calculate to get the *quotient* and *remainder* values. And from the *quotient* and *remainder* values, *Mathematical Calculation System* selects the maximum one and displays it on the screen. Figure 10-2 shows the *Dividend* value is 88 and the *Divisor* value is 5 and the calculated results of *quotient* and *remainder* respectively are 17 and 3. The maximum value of 17 and 3, which is 17, is then displayed on the screen.

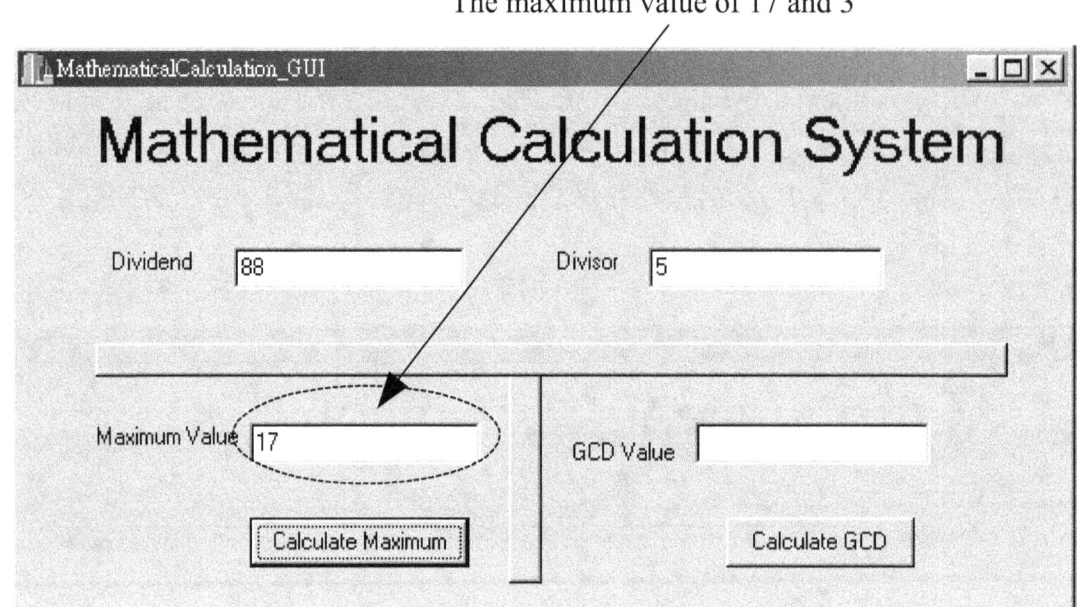

Figure 10-2   DIVIDE&MAXIMUM Behavior

In the *DIVIDE&GCD* behavior, the external actor inputs the *Dividend* and *Divisor* integer values then presses down the *Calculate_GCD* button. After that, *Mathematical Calculation System* will calculate to get the *quotient* and *remainder* values. And from the *quotient* and *remainder* values, *Mathematical Calculation System* will calculate the greatest common divisor value and displays it on the screen. Figure 10-3 shows the *Dividend* value is 224 and the *Divisor* value is 18 and the calculated results of *quotient* and *remainder* respectively are 12 and 8. The greatest common divisor value of 12 and 8, which is 4, is then displayed on the screen.

The GCD value of 12 and 8

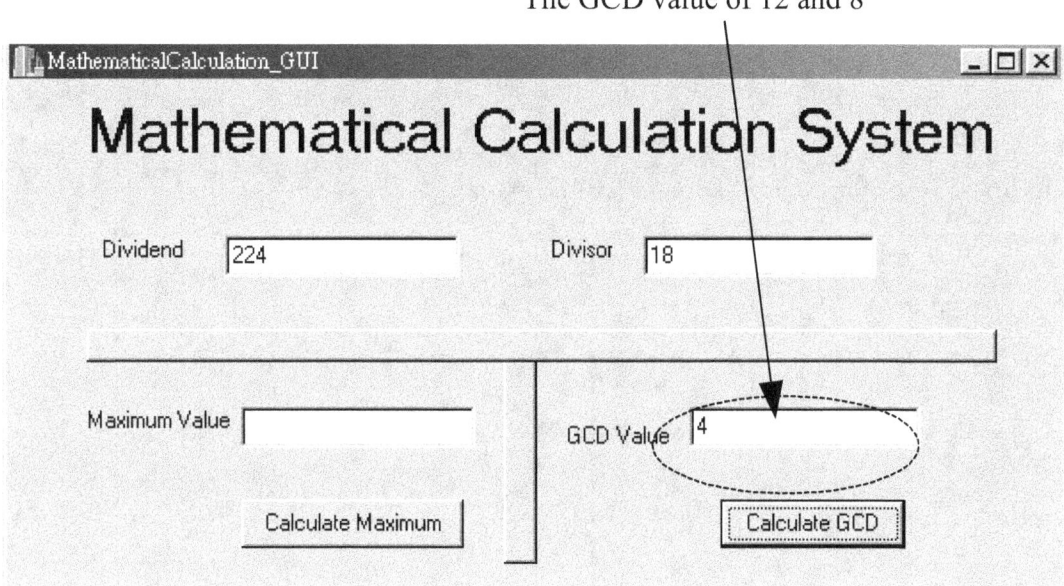

Figure 10-3   DIVIDE&GCD Behavior

In this chapter, we use the operation-based infinite-queue SBC process algebra to model the *Mathematical Calculation System* as shown in Figure 10-4.

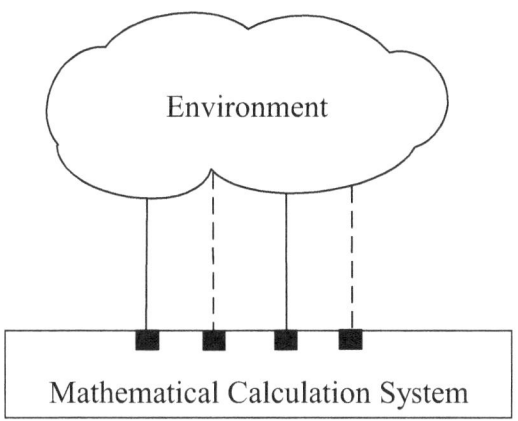

Figure 10-4   Systems Modeling of
the *Mathematical Calculation System*

10-1 BNF Tree of the Mathematical Calculation System

The operation-based infinite-queue SBC process of the *Mathematical Calculation System*, A_{031}, is defined as "$!(t_{031} \bullet t_{032} \bullet t_{033} \bullet t_{034} \bullet STOP) \| !(t_{035} \bullet t_{032} \bullet t_{037} \bullet t_{038} \bullet STOP)$".

We draw the operation-based infinite-queue SBC process algebra Backus-Naur Form tree of A_{031} as shown in Figure 10-5.

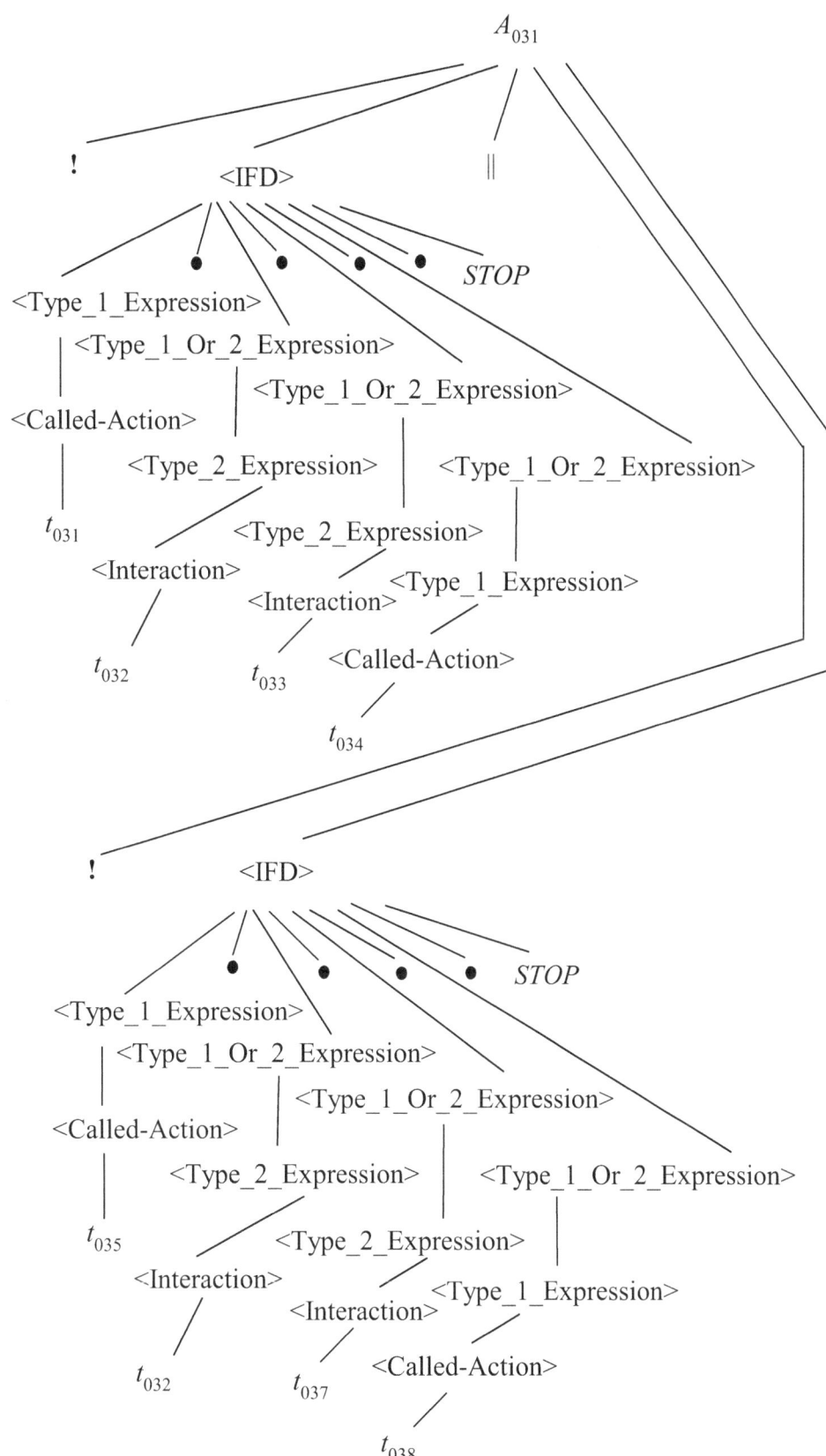

Figure 10-5   Backus-Naur Form Tree of the *Mathematical Calculation System*'s
Operation-Based Infinite-Queue SBC Process

There are two IFDs in the operation-based infinite-queue SBC process of the *Mathematical Calculation System*. The first IFD is shown in Figure 10-6.

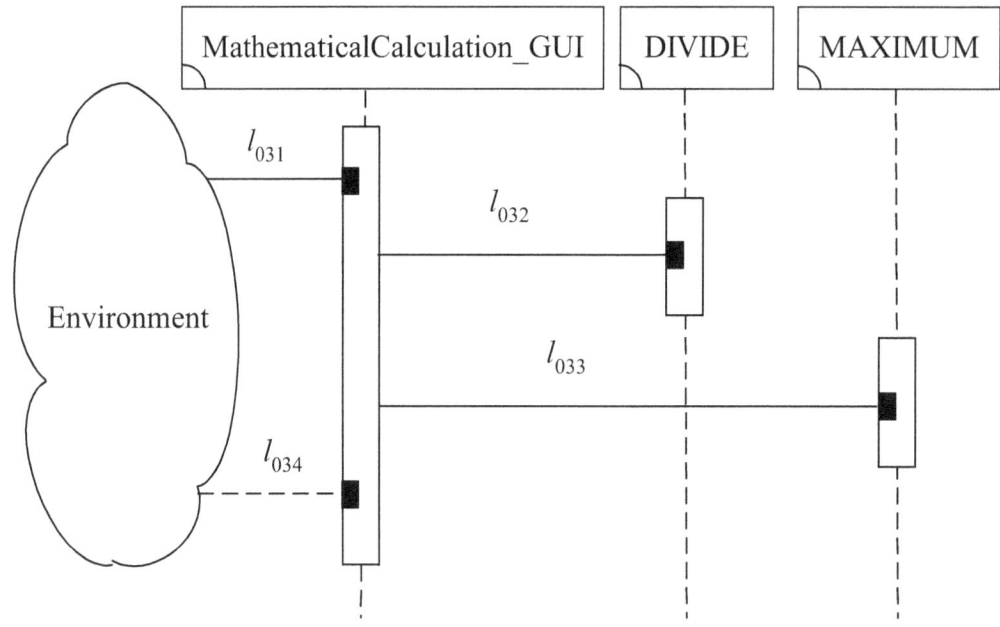

Figure 10-6   First IFD of the *Mathematical Calculation System*

The second IFD of the operation-based infinite-queue SBC process of the *Mathematical Calculation System* is shown in Figure 10-7.

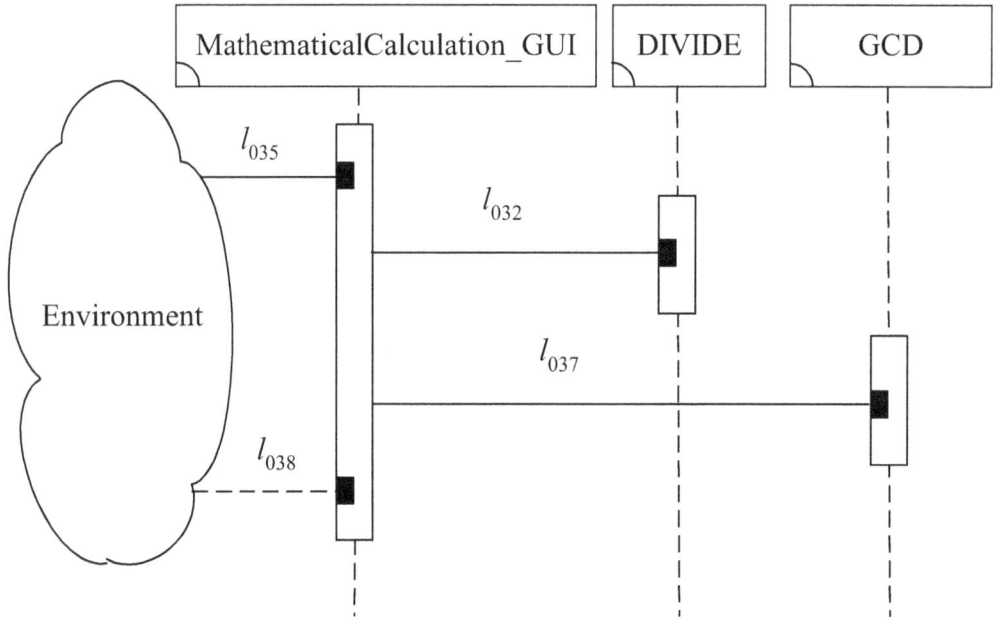

Figure 10-7   Second IFD of the *Mathematical Calculation System*

10-2 Prefixes of the Mathematical Calculation System

The t_{031} (operation-based called action with no conditions and no variable settings) prefix is defined as "<OPERATION_CALL, *MathematicalCalculation_GUI*, CALLED, l_{031}>", as shown in Figure 10-8.

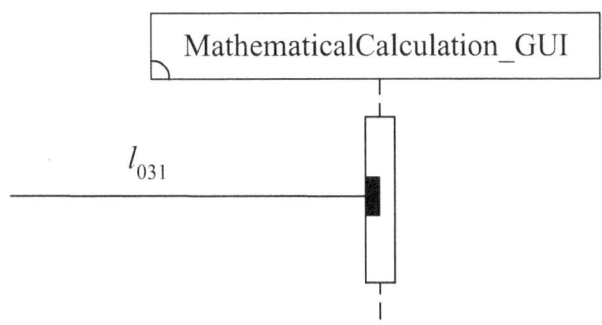

Figure 10-8   Prefix of t_{031}

The t_{032} (operation-based interaction with no conditions and no variable settings) prefix is defined as "<OPERATION_CALL, *MathematicalCalculation_GUI*, l_{032}, *DIVIDE*>", as shown in Figure 10-9.

Figure 10-9   Prefix of t_{032}

The t_{033} (operation-based interaction with no conditions and no variable settings) prefix is defined as "<OPERATION_CALL, *DIVIDE*, l_{033}, *MAXIMUM*>", as shown in Figure 10-10.

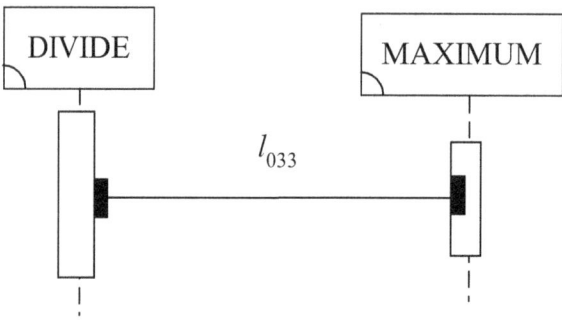

Figure 10-10   Prefix of t_{033}

The t_{034} (operation-based called action with no conditions and no variable settings) prefix is defined as "<OPERATION_RETURN, *MathematicalCalculation_GUI*, CALLED, l_{034}>", as shown in Figure 10-11.

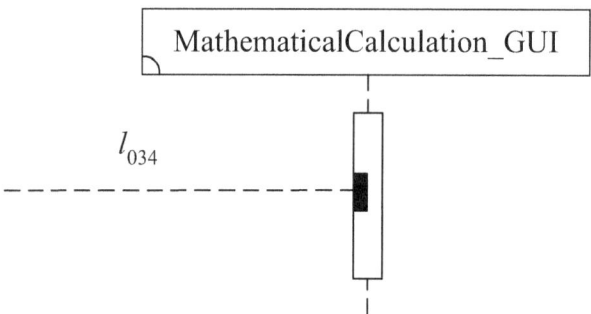

Figure 10-11   Prefix of t_{034}

The t_{035} (operation-based called action with no conditions and no variable settings) prefix is defined as "<OPERATION_CALL, *MathematicalCalculation_GUI*, CALLED, l_{035}>", as shown in Figure 10-12.

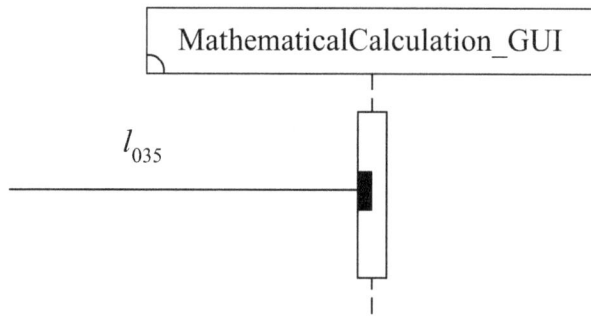

Figure 10-12   Prefix of t_{035}

The t_{037} (operation-based interaction with no conditions and no variable settings) prefix is defined as "<OPERATION_CALL, *DIVIDE*, l_{037}, *GCD*>", as shown in Figure 10-13.

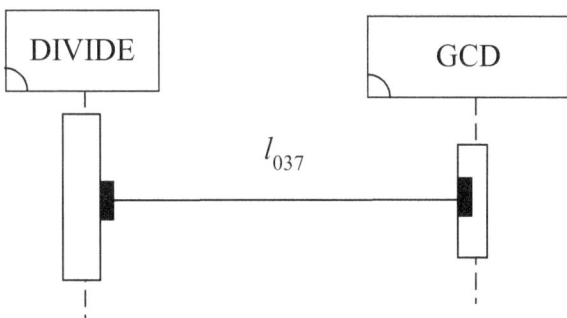

Figure 10-13   Prefix of t_{037}

The t_{038} (operation-based called action with no conditions and no variable settings) prefix is defined as "<OPERATION_RETURN, *MathematicalCalculation_GUI*, CALLED, l_{038}>", as shown in Figure 10-14.

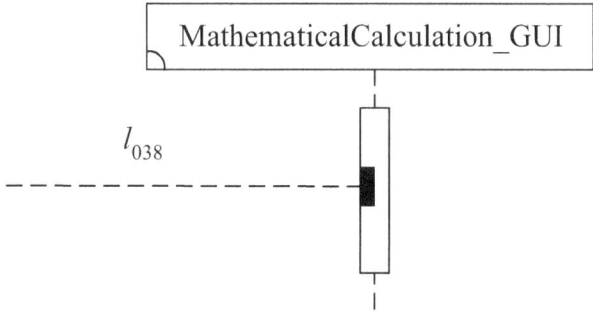

Figure 10-14   Prefix of t_{038}

Figure 10-15 shows all operation call or operation return formulas of the operation-based infinite-queue SBC process of the *Mathematical Calculation System*.

Entity name	Operation Call Formula	Operation Return Formula
l_{031}	Calculate_MaximumClick(In Dividend, Divisor)	
l_{032}	DIVIDE(In dividend, divisor; Out quotient, remainder)	
l_{033}	MAXIMUM(In quotient, remainder; Out maximum_value)	
l_{034}		Calculate_MaximumClick(Out Maximum_value)
l_{035}	Calculate_GCDClick(In Dividend, Divisor)	
l_{037}	GCD(In quotient, remainder; Out gcd_value)	
l_{038}		Calculate_GCDClick(Out GCD_value)

Figure 10-15   Operation Call or Operation Return Formulas
of the *Mathematical Calculation System*'s Process

Figure 10-16 shows the primitive data type specification of the *Dividend*, *Divisor*, *Maximum_value*, *GCD_value*, *dividend*, *divisor*, *quotient*, *remain*, *maximum_value* and *gcd_value* parameters.

Parameter	Data Type	Instances
Dividend	Integer	88, 188
Divisor	Integer	5, 15
Maximum_value	Integer	17, 12
GCD_value	Integer	1, 4
dividend	Integer	88, 188
divisor	Integer	5, 15
quotient	Integer	17, 12
remain	Integer	3, 8
maximum_value	Integer	17, 12
gcd_value	Integer	1, 4

Figure 10-16　　Primitive Data Type Specification

10-3 Process of the Mathematical Calculation System

The following transition graph shows, in Figure 10-17, the semantics of P_{031}'s operation-based infinite-queue SBC process.

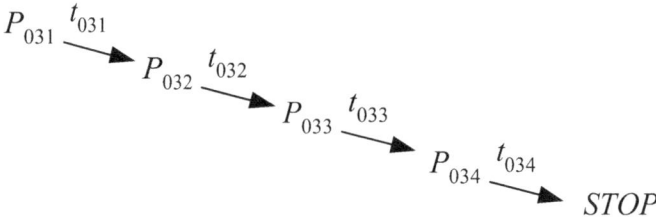

Figure 10-17   Transition graph of the P_{031} Process

In the transition graph of the P_{031}'s operation-based infinite-queue SBC process, processes P_{031}, P_{032}, P_{033} and P_{034} are defined as in Figure 10-18.

$$P_{031} \overset{\text{def}}{=\!=} t_{031} \bullet P_{032}$$

$$P_{032} \overset{\text{def}}{=\!=} t_{032} \bullet P_{033}$$

$$P_{033} \overset{\text{def}}{=\!=} t_{033} \bullet P_{034}$$

$$P_{034} \overset{\text{def}}{=\!=} t_{034} \bullet STOP$$

Figure 10-18   Definition of Processes
$P_{031}, P_{032}, P_{033},$ and P_{034}

The following transition graph shows, in Figure 10-19, the semantics of P_{035}'s operation-based infinite-queue SBC process.

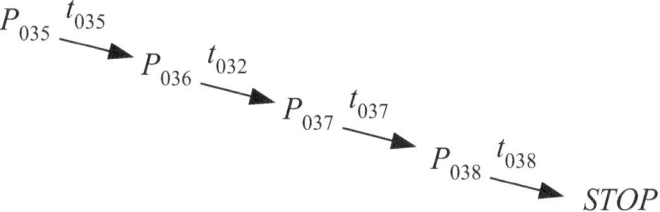

Figure 10-19   Transition graph of the P_{035} Process

In the transition graph of the P_{035}'s operation-based infinite-queue SBC process, processes P_{035}, P_{036}, P_{037} and P_{038} are defined as in Figure 10-20.

$$P_{035} \overset{\text{def}}{=\joinrel=} t_{035} \bullet P_{036}$$

$$P_{036} \overset{\text{def}}{=\joinrel=} t_{032} \bullet P_{037}$$

$$P_{037} \overset{\text{def}}{=\joinrel=} t_{037} \bullet P_{038}$$

$$P_{038} \overset{\text{def}}{=\joinrel=} t_{038} \bullet STOP$$

Figure 10-20   Definition of Processes
P_{035}, P_{036}, P_{037}, and P_{038}

The operation-based infinite-queue SBC process of the *Multi-Tier Personal Data System*, A_{031}, is defined as "$!P_{031} \| !P_{035}$".

Chapter 11: Operation-Based Infinite-Queue SBC Process of the Web Service Arithmetic System

After the system development is completed, the *Web Service Arithmetic System* shall appear on a web multi-tier platform [Sebe12] as shown in Figure 11-1.

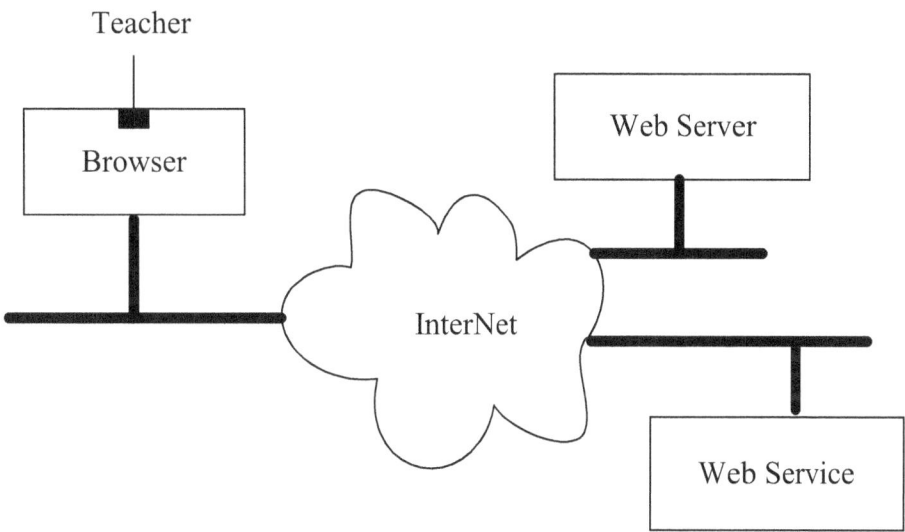

Figure 11-1 *Web Service Arithmetic System* on a Web Multi-Tier Platform

The functionality of the *Web Service Arithmetic System* is to provide a web browser for the *Teacher* actor to input the P, Q, R, S and T values as shown in Figure 11-2.

Input the *P, Q, R, S, T* values

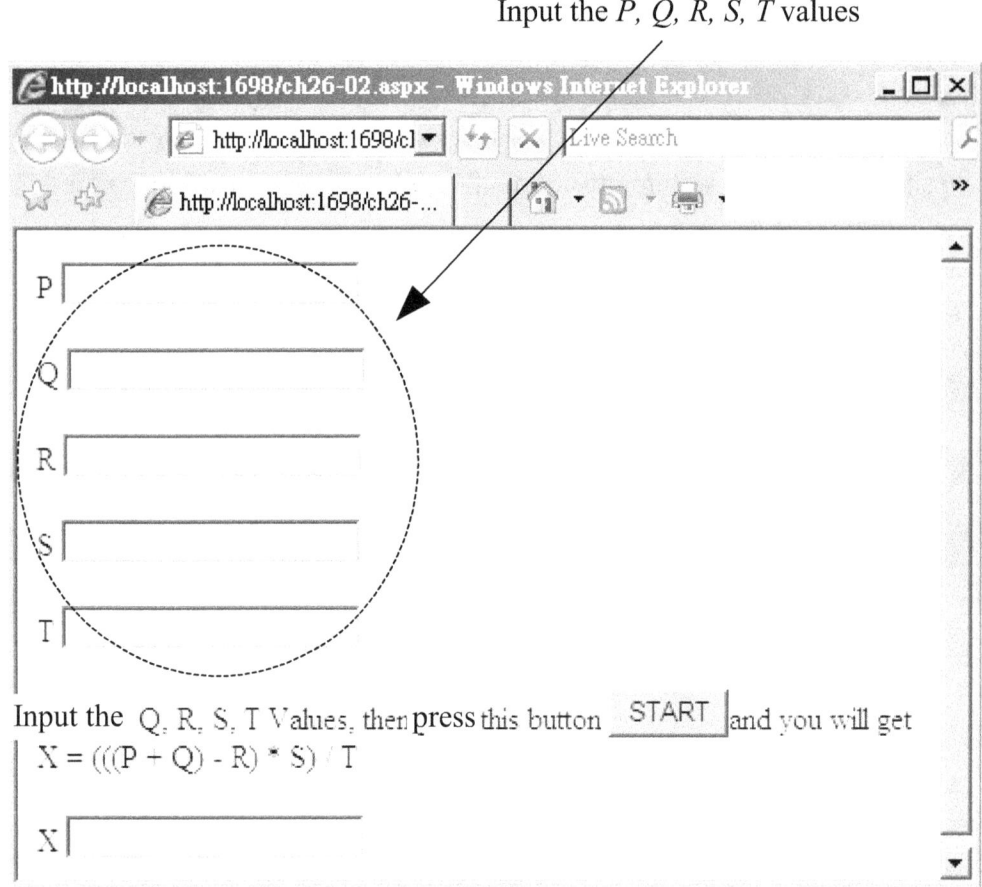

Figure 11-2   Web Browser for the *Pupil* Actor to Input the Values

When the *START* button is pressed, the *Web Service Arithmetic System* calculates the (((P+Q)-R) *S)/T value and presents the result on the *X* output box, as shown in Figure 11-3.

Figure 11-3   Result after Pressing Down the *START* Button

In this chapter, we use the operation-based infinite-queue SBC process algebra to model the *Web Service Arithmetic System* as shown in Figure 11-4.

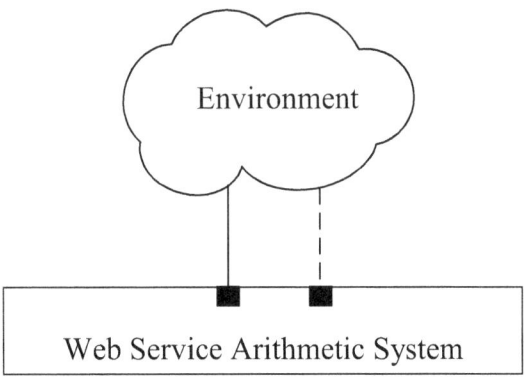

Figure 11-4   Systems Modeling
the *Web Service Arithmetic System*

11-1 BNF Tree of the Web Service Arithmetic System

The operation-based infinite-queue SBC process of the *Web Service Arithmetic System*, A_{041}, is defined as "$!(t_{041} \bullet t_{042} \bullet t_{043} \bullet t_{044} \bullet t_{045} \bullet t_{046} \bullet STOP)$".

We draw the operation-based infinite-queue SBC process algebra Backus-Naur Form tree of A_{041} as shown in Figure 11-5.

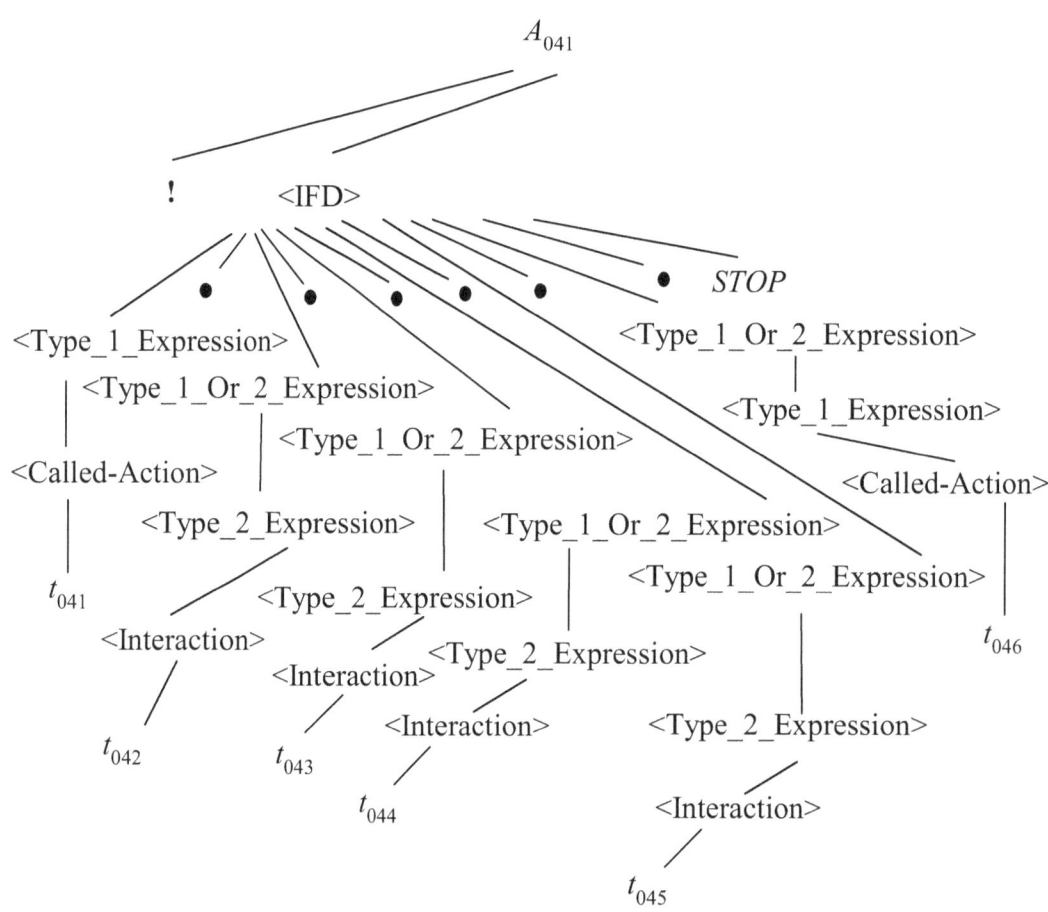

Figure 11-5   Backus-Naur Form Tree of the *Web Service Arithmetic System*'s
Operation-Based Infinite-Queue SBC Process

There is only one IFD in the operation-based infinite-queue SBC process of the *Web Service Arithmetic System*. The only IFD is shown in Figure 11-6.

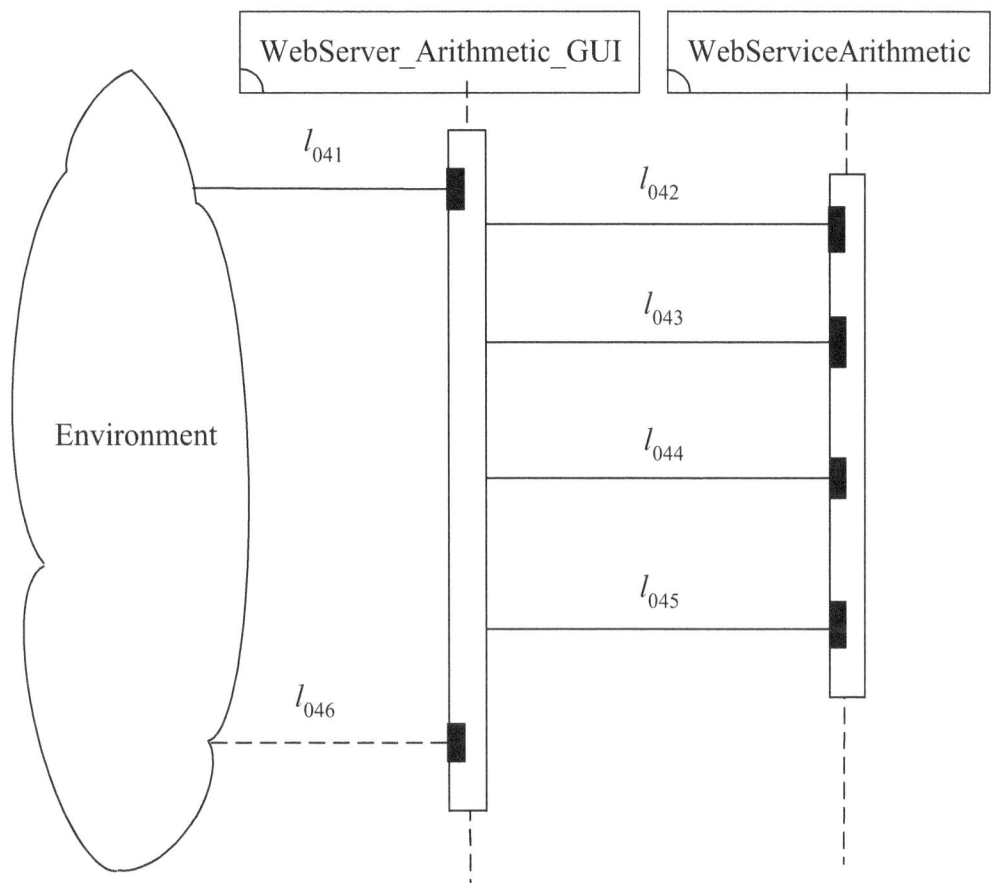

Figure 11-6   The Only IFD of the *Web Service Arithmetic System*

11-2 Prefixes of the Web Service Arithmetic System

The t_{041} (operation-based called action with no conditions and no variable settings) prefix is defined as "<OPERATION_CALL, *WebServer_Arithmetic_GUI*, CALLED, l_{041}>", as shown in Figure 11-7.

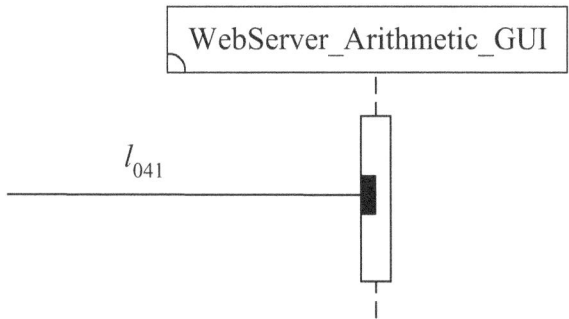

Figure 11-7   Prefix of t_{041}

The t_{042} (operation-based interaction with no conditions and no variable settings) prefix is defined as "<OPERATION_CALL, *WebServer_Arithmetic_GUI*, l_{042}, *WebServiceArithmetic*>", as shown in Figure 11-8.

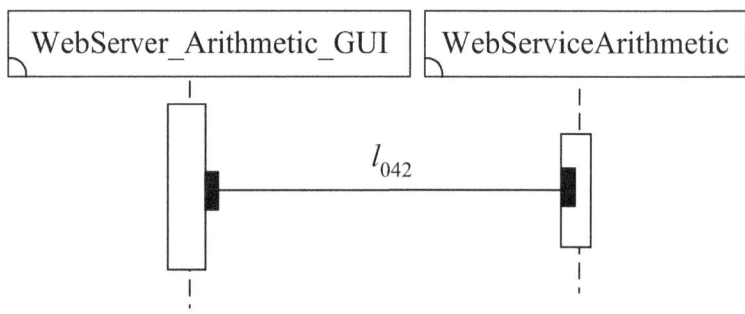

Figure 11-8   Prefix of t_{042}

The t_{043} (operation-based interaction with no conditions and no variable settings) prefix is defined as "<OPERATION_CALL, *WebServer_Arithmetic_GUI*, l_{043}, *WebServiceArithmetic*>", as shown in Figure 11-9.

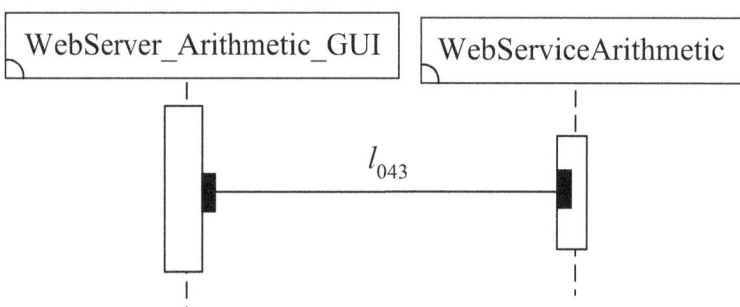

Figure 11-9   Prefix of t_{043}

The t_{044} (operation-based interaction with no conditions and no variable settings) prefix is defined as "<OPERATION_CALL, *WebServer_Arithmetic_GUI*, l_{044}, *WebServiceArithmetic*>", as shown in Figure 11-10.

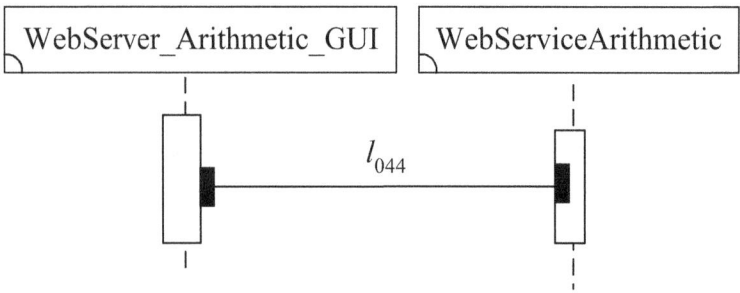

Figure 11-10   Prefix of t_{044}

The t_{045} (operation-based interaction with no conditions and no variable settings) prefix is defined as "<OPERATION_CALL, *WebServer_Arithmetic_GUI*, l_{045}, *WebServiceArithmetic*>", as shown in Figure 11-11.

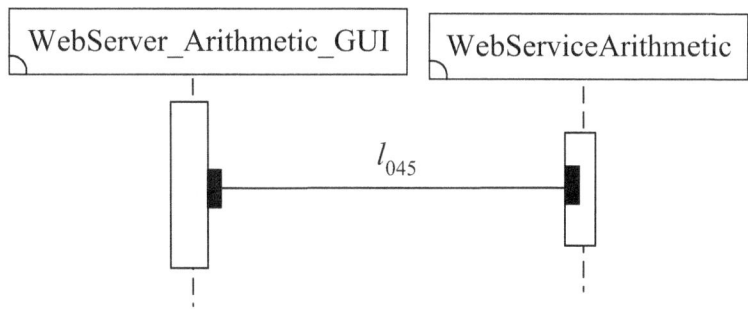

Figure 11-11   Prefix of t_{045}

The t_{046} (operation-based called action with no conditions and no variable settings) prefix is defined as "<OPERATION_RETURN, *WebServer_Arithmetic_GUI*, CALLED, l_{046}>", as shown in Figure 11-12.

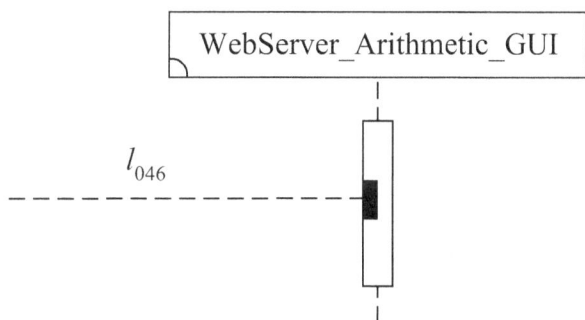

Figure 11-12   Prefix of t_{046}

Figure 11-13 shows all operation call or operation return formulas of the operation-based infinite-queue SBC process of the *Web Service Arithmetic System.*

Entity name	Operation Call Formula	Operation Return Formula
l_{041}	START_Click(In P, Q, R, S, T)	
l_{042}	Add(In a, b; Out c)	
l_{043}	Subtract(In d, e; Out f)	
l_{044}	Multiply(In g, h; Out i)	
l_{045}	Divide(In j, k; Out m)	
l_{046}		START_Click(Out X)

Figure 11-13   Operation Call or Operation Return Formulas
of the *Web Service Arithmetic System*'s Process

Figure 11-14 shows the primitive data type specification of the *P, Q, R, S, T, a, b, d, e, g, h, j, k* input parameter and the *X, c, f, i, m* output parameters.

Parameter	Data Type	Instances
P	Integer	20, 7
Q	Integer	2, 6
R	Integer	8, 5
S	Integer	5, 4
T	Integer	3, 3
X	Integer	23, 11
a	Integer	20, 7
b	Integer	2, 6
c	Integer	22, 13
d	Integer	22, 13
e	Integer	8, 5
f	Integer	14, 8
g	Integer	14, 8
h	Integer	5, 4
i	Integer	70, 32
j	Integer	70, 32
k	Integer	3, 3
m	Integer	23, 11

Figure 11-14   Primitive Data Type Specification

104

11-3 Process of the Web Service Arithmetic System

The following transition graph shows, in Figure 11-15, the semantics of A_{041}'s operation-based infinite-queue SBC process.

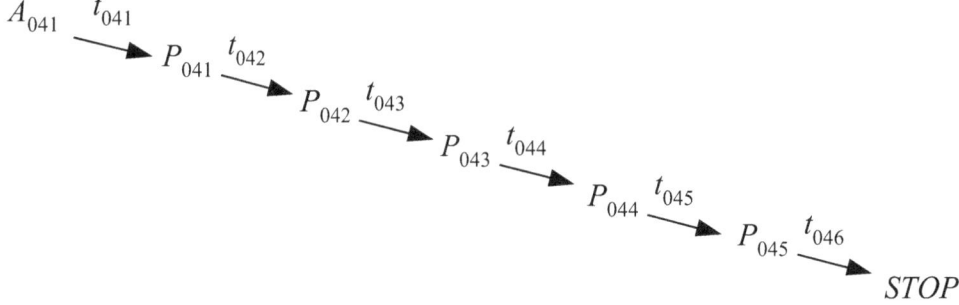

Figure 11-15   Transition graph
of the *Web Service Arithmetic System*'s Process

In the transition graph of the A_{041}'s operation-based infinite-queue SBC process, processes A_{041}, P_{041}, P_{042}, P_{043}, P_{044} and P_{045} are defined as in Figure 11-16.

$$A_{041} \stackrel{\text{def}}{=\!=} t_{041} \bullet P_{041}$$

$$P_{041} \stackrel{\text{def}}{=\!=} t_{042} \bullet P_{042}$$

$$P_{042} \stackrel{\text{def}}{=\!=} t_{043} \bullet P_{043}$$

$$P_{043} \stackrel{\text{def}}{=\!=} t_{044} \bullet P_{044}$$

$$P_{044} \stackrel{\text{def}}{=\!=} t_{045} \bullet P_{045}$$

$$P_{045} \stackrel{\text{def}}{=\!=} t_{046} \bullet STOP$$

Figure 11-16   Definition of Processes
A_{041}, P_{041}, P_{042}, P_{043}, P_{044}, and P_{045}

Chapter 12: Operation-Based Infinite-Queue SBC Process of the Web Service Extranet System

After the system development is completed, the *web service extranet system* shall appear on a web multi-tier platform [Sebe12] as shown in Figure 12-1.

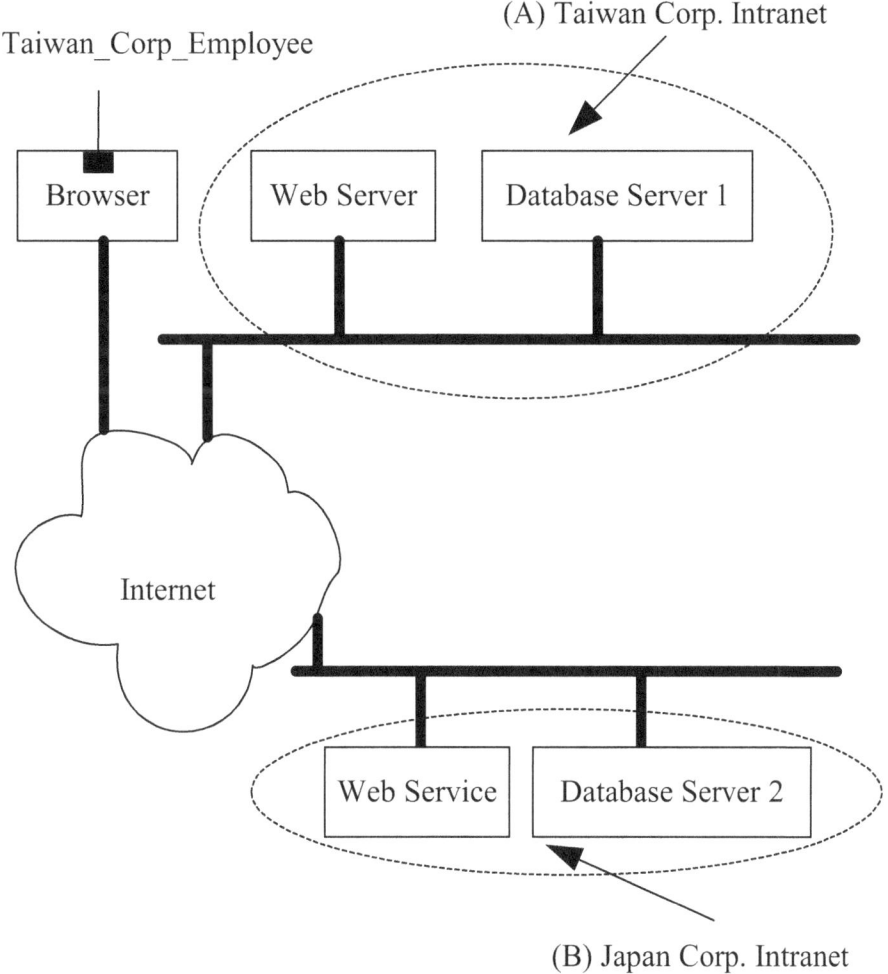

Figure 12-1   *Web Service Extranet System* on a Web Multi-Tier Platform

The major functionality of the *Web Service Extranet System* is to provide a web browser for the *Taiwan_Corp_Employee* actor to input the *PurchaseDate* value as shown in Figure 12-2.

106

Input the Purchase Date

Figure 12-2   Web Browser for the Actor to input the *Purchase Date*

After the *Taiwan_Corp_Employee* actor inputs the *PurchaseDate* value and presses down the *ExtranetButton* button then the *Web Service Extranet System* generates the following two results:

(A) The first result is to insert a new record into the *PurchaseTable* table of the *Taiwan_Database* database as shown in Figure 12-3.

PurchaseDate	PurchaseNo	Suppliename
20070425	001	Japan's Corp.
20080817	001	Japan's Corp.
20090312	001	Japan's Corp.

Record: 14 4   3 ▶ ▶1 ▶* of 3

Inserting a new record

Figure 12-3   First Result After Pressing Down the *ExtranetButton* Button

(B) The second result is to insert a new record into the *SaleTable* table of the *Japan_Database* database as shown in Figure 12-4.

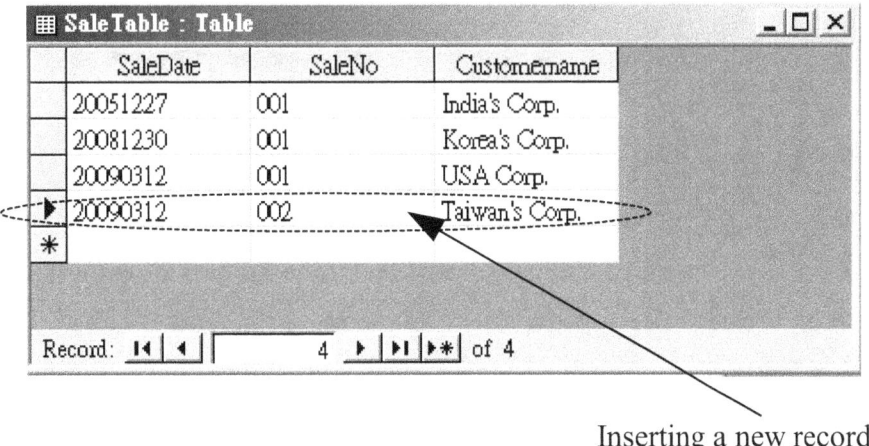

Inserting a new record

Figure 12-4   Second Result After Pressing Down the *ExtranetButton* Button

Finally, we also see a message informing us that the Extranet deal is done, on the web page, as shown in Figure 12-5.

ExtraNet is Done

Figure 12-5   Informing that the Extranet Deal is Done

In this chapter, we use the operation-based infinite-queue SBC process algebra to model the *Web Service Extranet System* as shown in Figure 12-6.

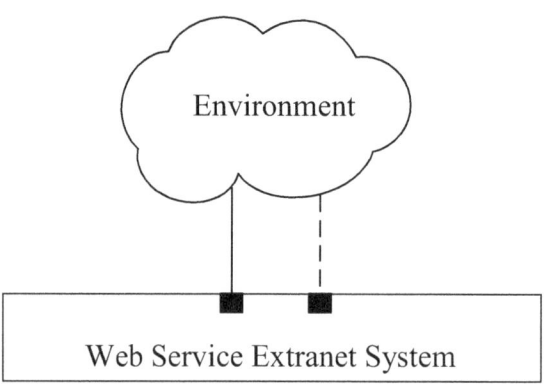

Figure 12-6   Systems Modeling
the *Web Service Extranet System*

12-1 BNF Tree of the Web Service Extranet System

The operation-based infinite-queue SBC process of the *Web Service Extranet System*, A_{081}, is defined as "$!(t_{081} \bullet t_{082} \bullet t_{083} \bullet t_{084} \bullet t_{085} \bullet STOP)$".

We draw the operation-based infinite-queue SBC process algebra Backus-Naur Form tree of A_{081} as shown in Figure 12-7.

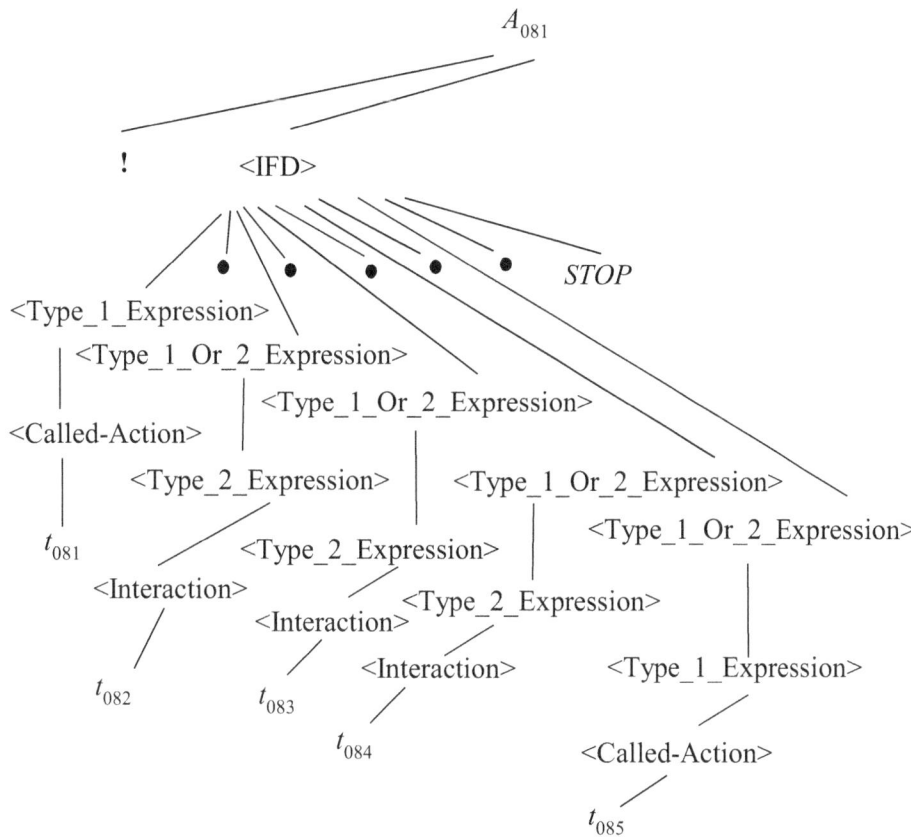

Figure 12-7   Backus-Naur Form Tree of the *Web Service Extranet System*'s
Operation-Based Infinite-Queue SBC Process

There is only one IFD in the operation-based infinite-queue SBC process of the *Web Service Extranet System*. The only IFD is shown in Figure 12-8.

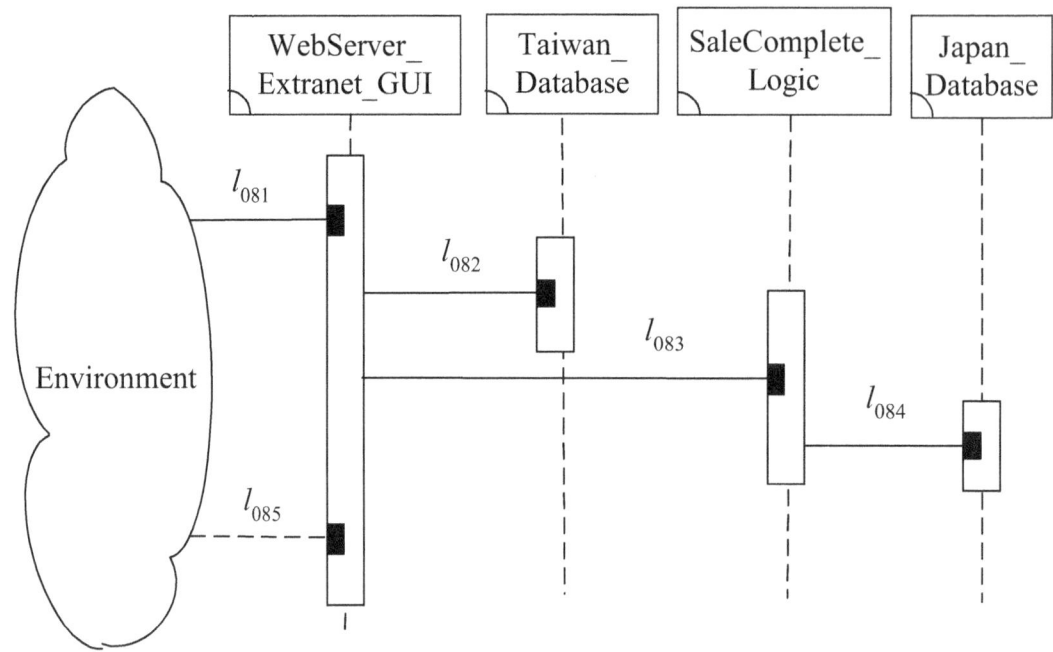

Figure 12-8　IFD of the *Web Service Extranet System*

12-2 Prefixes of the Web Service Extranet System

The t_{081} (operation-based called action with no conditions and no variable settings) prefix is defined as "<OPERATION_CALL, *WebServer_Extranet_GUI*, CALLED, l_{081}>", as shown in Figure 12-9.

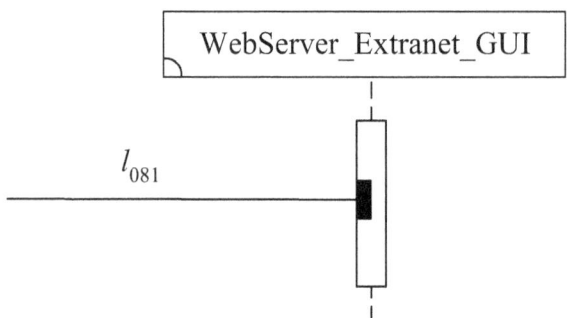

Figure 12-9　Prefix of t_{081}

The t_{082} (operation-based interaction with no conditions and no variable settings) prefix is defined as "<OPERATION_CALL, *WebServer_Extranet_GUI*, l_{082}, *Taiwan_Database*>", as shown in Figure 12-10.

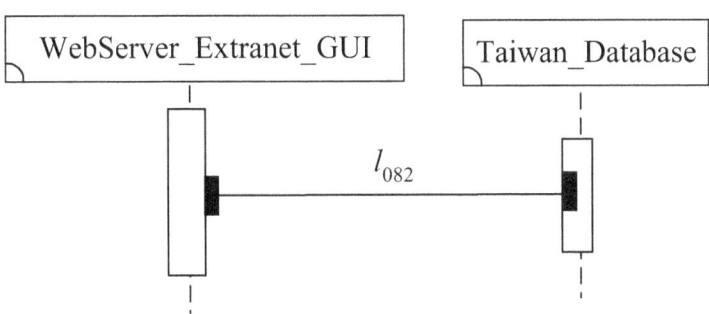

Figure 12-10   Prefix of t_{082}

The t_{083} (operation-based interaction with no conditions and no variable settings) prefix is defined as "<OPERATION_CALL, *WebServer_Extranet_GUI*, l_{083}, *SaleComplete_Logic*>", as shown in Figure 12-11.

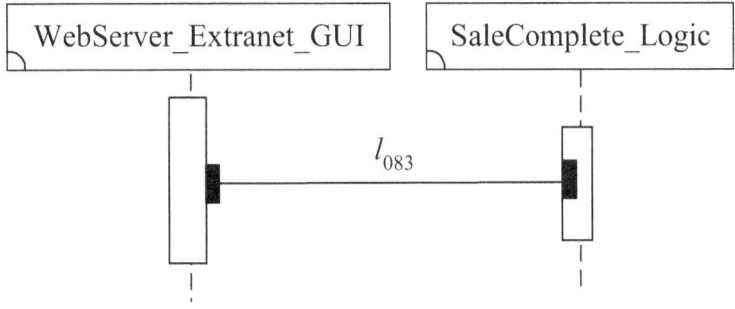

Figure 12-11   Prefix of t_{083}

The t_{084} (operation-based interaction with no conditions and no variable settings) prefix is defined as "<OPERATION_CALL, *SaleComplete_Logic*, l_{084}, *Japan_Database*>", as shown in Figure 12-12.

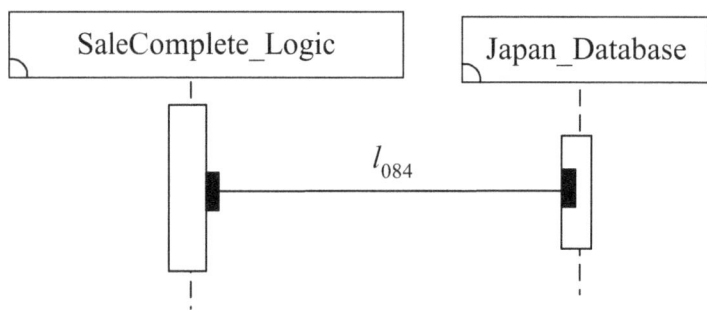

Figure 12-12   Prefix of t_{084}

The t_{085} (operation-based called action with no conditions and no variable settings) prefix is defined as "<OPERATION_RETURN, *WebServer_Extranet_GUI*, CALLED, l_{085}>", as shown in Figure 12-13.

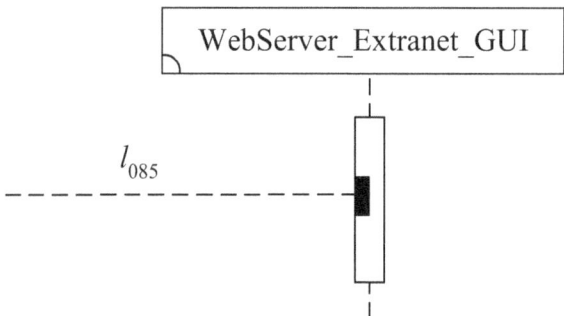

Figure 12-13   Prefix of t_{085}

Figure 12-14 shows all operation call or operation return formulas of the operation-based infinite-queue SBC process of the *Web Service Extranet System.*

Entity name	Operation Call Formula	Operation Return Formula
l_{081}	ExtranetButton_Click(In PurchaseDate)	
l_{082}	Add_a_Sale(In SaleDate, Customer)	
l_{083}	Sql_p_Insert is Sql_p_Insert(In p_query)	
l_{084}	Sql_s_Insert is Sql_s_Insert(In s_query)	
l_{085}		ExtranetButton_Click

Figure 12-14   Operation Call or Operation Return Formulas
of the *Web Service Extranet System*'s Process

Figure 12-15 shows the primitive data type specification of the *PurchaseDate, SaleDate* and *Customer* input parameters.

Parameter	Data Type	Instances
PurchaseDate	String	20090112
SaleDate	String	20090112
Customer	String	Taiwan Corp.

Figure 12-15   Primitive Data Type Specification

Figure 12-16 shows the composite data type specification of the *p_query* input parameter occurring in the *Sql_p_Insert(In p_query)* operation call formula.

Parameter	*p_query*		
Data Type	TABLE of Purchase Date : Text Purchase No : Text Supplier : Text End TABLE ;		
Instances	PurchaseDate	PurchaseNo	Supplier
	20090312	001	Japan Corp.

Figure 12-16   Composite Data Type Specification

Figure 12-17 shows the composite data type specification of the *s_query* input parameter occurring in the *Sql_s_Insert(In s_query)* operation call formula.

Parameter	*s_query*		
Data Type	TABLE of SaleDate : Text SaleNo : Text Customer : Text End TABLE ;		
Instances	Sale Date	Sale No	Customer
	20090312	001	Taiwan Corp.

Figure 12-17   Composite Data Type Specification

12-3 Process of the Web Service Extranet System

The following transition graph shows, in Figure 12-18, the semantics of A_{081}'s operation-based infinite-queue SBC process.

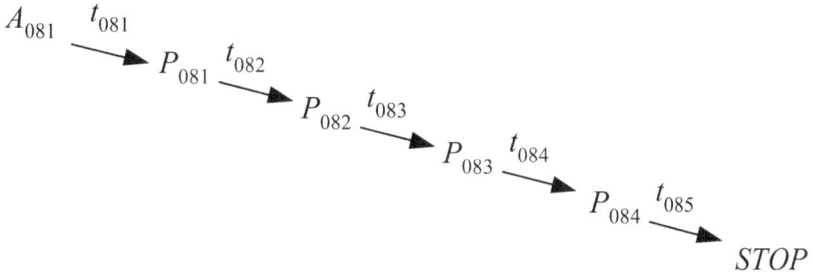

Figure 12-18   Transition graph
of the *Web Service Extranet System*'s Process

In the transition graph of the A_{081}'s operation-based infinite-queue SBC process, processes A_{081}, P_{081}, P_{082}, P_{083} and P_{084} are defined as in Figure 12-19.

$$A_{081} \overset{\text{def}}{=\joinrel=} t_{081} \bullet P_{081}$$

$$P_{081} \overset{\text{def}}{=\joinrel=} t_{082} \bullet P_{082}$$

$$P_{082} \overset{\text{def}}{=\joinrel=} t_{083} \bullet P_{083}$$

$$P_{083} \overset{\text{def}}{=\joinrel=} t_{084} \bullet P_{084}$$

$$P_{084} \overset{\text{def}}{=\joinrel=} t_{085} \bullet STOP$$

Figure 12-19   Definition of Processes
A_{081}, P_{081}, P_{082}, P_{083}, and P_{084}

Chapter 13: Operation-Based Infinite-Queue SBC Process of the Convenience Store's Get 2nd 50% off Sales Promotion System

In order to attract customers to buy, a convenience store may find many ways to do so. The easiest one is to give discounts. Because of human nature, coupled with the hot summer weather, the customers see the advertisement saying a cold drink at 50% off, they may have a great desire to buy it. This inspires the idea of a *Convenience Store's Get 2nd 50% off Sales Promotion System*.

In this chapter, we use the operation-based infinite-queue SBC process algebra to model the *Convenience Store's Get 2nd 50% off Sales Promotion System* as shown in Figure 13-1.

Figure 13-1   Systems Modeling the *Convenience Store's Get 2nd 50% off Sales Promotion System*

13-1 BNF Tree of the Convenience Store's Get 2nd 50% off Sales Promotion System

The operation-based infinite-queue SBC process of the *Convenience Store's Get 2nd 50% off Sales Promotion System*, A_{101}, is defined as "$!(t_{101} \bullet STOP)\|!(t_{102} \bullet t_{103} \bullet STOP)$".

We draw the operation-based infinite-queue SBC process algebra Backus-Naur Form tree of A_{101} as shown in Figure 13-2.

118

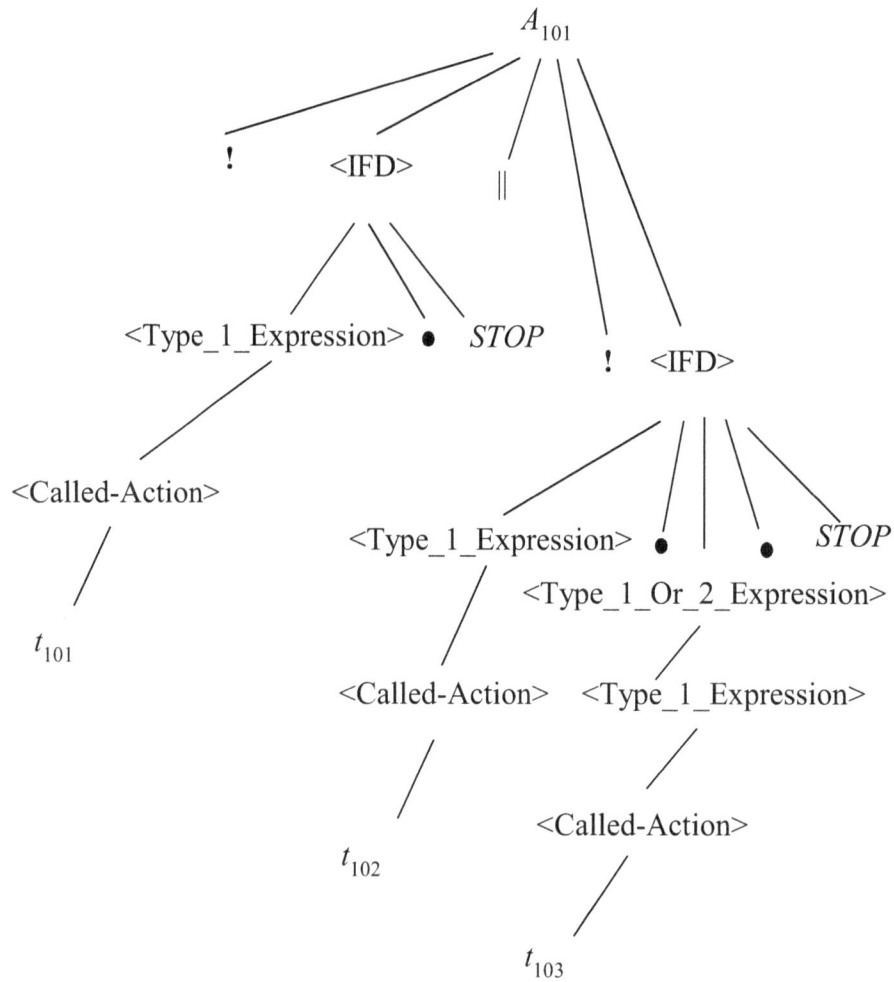

Figure 13-2 Backus-Naur Form Tree of the *Convenience Store's Get 2nd 50% off Sales Promotion System*'s Operation-Based Infinite-Queue SBC Process

There are two IFDs in the operation-based infinite-queue SBC process of the *Convenience Store's Get 2nd 50% off Sales Promotion System*. The first IFD is shown in Figure 13-3.

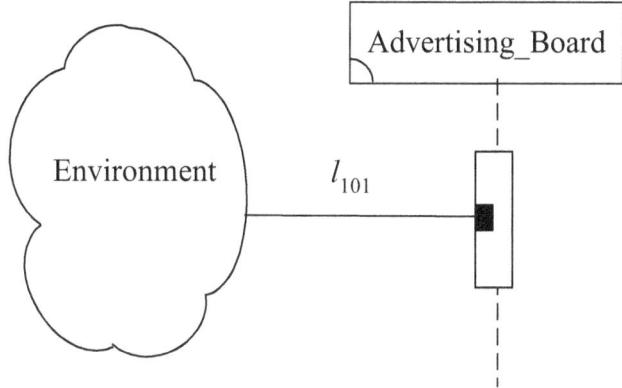

Figure 13-3   First IFD of the *Convenience Store's*
Get 2nd 50% off Sales Promotion System

The second IFD of the operation-based infinite-queue SBC process of the *Convenience Store's Get 2nd 50% off Sales Promotion System* is shown in Figure 13-4.

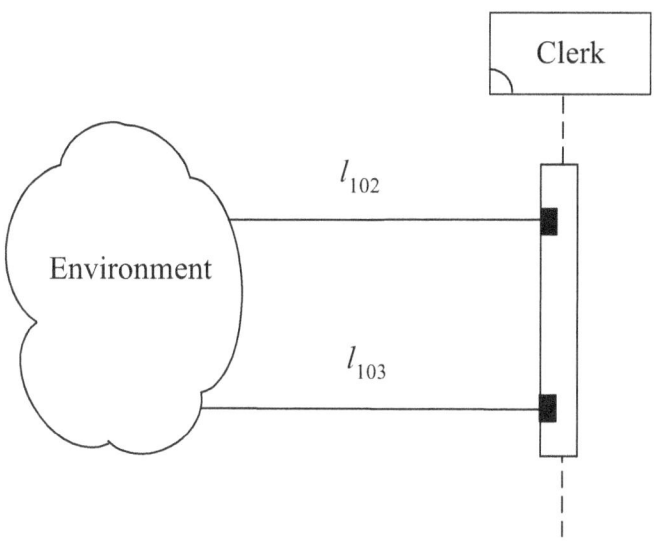

Figure 13-4   Second IFD of the *Convenience Store's*
Get 2nd 50% off Sales Promotion System

120

13-2 Prefixes of the Convenience Store's Get 2nd 50% off Sales Promotion System

The t_{101} (operation-based called action with no conditions and no variable settings) prefix is defined as "<OPERATION_CALL, *Advertising_Board*, CALLED, l_{101}>", as shown in Figure 13-5.

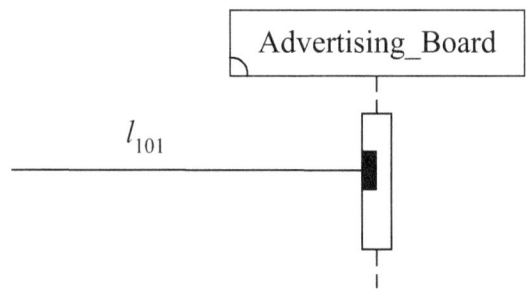

Figure 13-5   Prefix of t_{101}

The t_{102} (operation-based called action with no conditions and no variable settings) prefix is defined as "<OPERATION_CALL, *Clerk*, CALLED, l_{102}>", as shown in Figure 13-6.

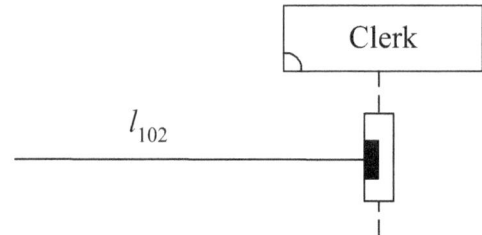

Figure 13-6   Prefix of t_{102}

The t_{103} (operation-based called action with no conditions and no variable settings) prefix is defined as "<OPERATION_CALL, *Clerk*, CALLED, l_{103}>", as shown in Figure 13-7.

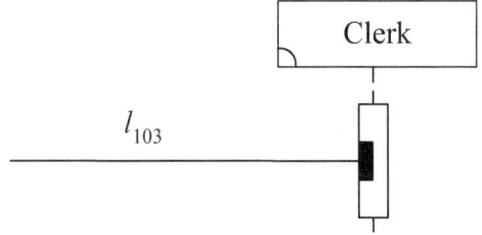

Figure 13-7   Prefix of t_{103}

Figure 13-8 shows all operation call or operation return formulas of the operation-based infinite-queue SBC process of the *Convenience Store's Get 2nd 50% off Sales Promotion System.*

Entity name	Operation Call Formula	Operation Return Formula
l_{101}	Watch(Out Advertisement)	
l_{102}	Buy_the_1st_Goods(In Payment; Out Goods)	
l_{103}	Buy_the_2nd_Goods(In 50%_off_Payment; Out Cold_Drink)	

Figure 13-8　Operation Call or Operation Return Formulas of the
Convenience Store's Get 2nd 50% off Sales Promotion System's Process

Figure 13-9 shows the primitive data type specification of the *Payment, 50%_off_Payment* input parameters and the *Advertisement, Goods, Cold_Drink* output parameters.

Parameter	Data Type	Instances
Payment	Money	$100
50%_off_Payment	Money	$50
Advertisement	Text	Welcome here!!!
Goods	Bulk	3, 9
Cold_Drink	Bottle	20, 30

Figure 13-9　Primitive Data Type Specification

13-3 Process of the Convenience Store's Get 2nd 50% off Sales Promotion System

The following transition graph shows, in Figure 13-10, the semantics of P_{101}'s operation-based infinite-queue SBC process.

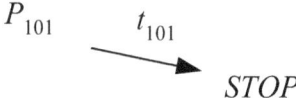

$$P_{101} \xrightarrow{t_{101}} STOP$$

Figure 13-10   Transition graph of the P_{101} Process

In the transition graph of the P_{101}'s operation-based infinite-queue SBC process, process P_{101} is defined as in Figure 13-11.

$$P_{101} \overset{\mathrm{def}}{=\!=} t_{101} \bullet STOP$$

Figure 13-11   Definition of Process P_{101}

The following transition graph shows, in Figure 13-12, the semantics of P_{102}'s operation-based infinite-queue SBC process.

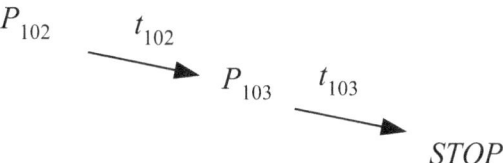

Figure 13-12   Transition graph of the P_{102} Process

In the transition graph of the P_{102}'s operation-based infinite-queue SBC process, processes P_{102} and P_{103} are defined as in Figure 13-13.

$$P_{102} \overset{\text{def}}{=\joinrel=} t_{102} \bullet P_{103}$$

$$P_{103} \overset{\text{def}}{=\joinrel=} t_{103} \bullet STOP$$

Figure 13-13   Definition of Processes P_{102} and P_{103}

The operation-based infinite-queue SBC process of the *Convenience Store's Get 2nd 50% off Sales Promotion System*, A_{101}, is defined as "$!P_{101}\|!P_{102}$".

Chapter 14: Operation-Based Infinite-Queue SBC Process of the Department Store's Car Sweepstakes Sales Promotion System

A decent car is something everyone wants to have. A customer will be interested in buying a thousand dollars goods in a department store if he has the opportunity to get a grand new one-million-dollars car. This inspires the idea of a *Department Store's Car Sweepstakes Sales Promotion System.*

In this chapter, we use the operation-based infinite-queue SBC process algebra to model the *Department Store's Car Sweepstakes Sales Promotion System* as shown in Figure 14-1.

Figure 14-1   Systems Modeling the *Department Store's Car Sweepstakes Sales Promotion System*

14-1 BNF Tree of the Department Store's Car Sweepstakes Sales Promotion System

The operation-based infinite-queue SBC process of the *Department Store's Car Sweepstakes Sales Promotion System*, A_{141}, is defined as "$!(t_{141} \bullet t_{142} \bullet STOP) \| !(t_{143} \bullet t_{144} \bullet t_{145} \bullet STOP) \| !(t_{146} \bullet STOP)$".

126

We draw the operation-based infinite-queue SBC process algebra Backus-Naur Form tree of A_{141} as shown in Figure 14-2.

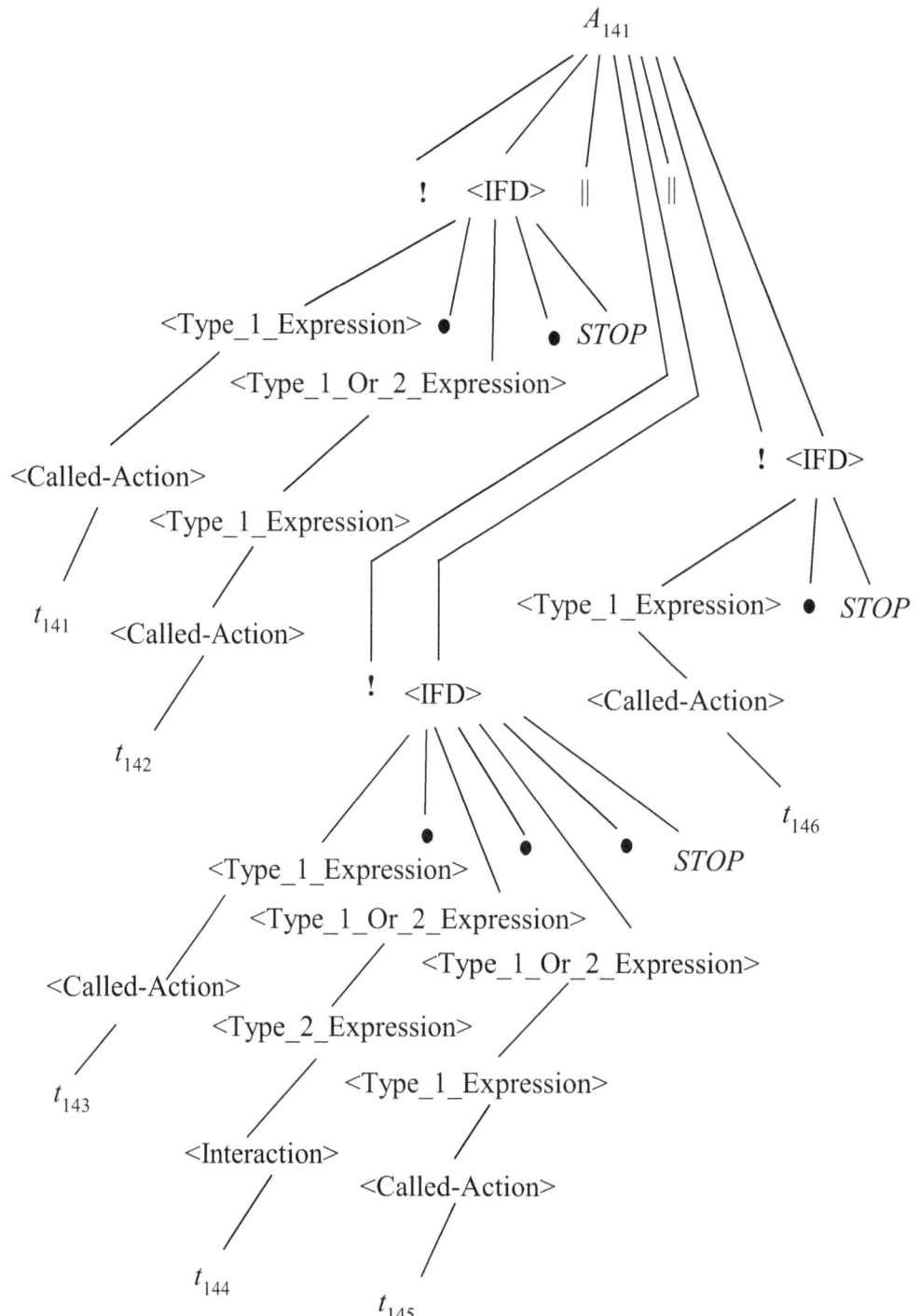

Figure 14-2   Backus-Naur Form Tree of the *Department Store's Car Sweepstakes Sales Promotion System*'s Operation-Based Infinite-Queue SBC Process

There are three IFDs in the operation-based infinite-queue SBC process of the *Department Store's Car Sweepstakes Sales Promotion System*. The first IFD is shown in Figure 14-3.

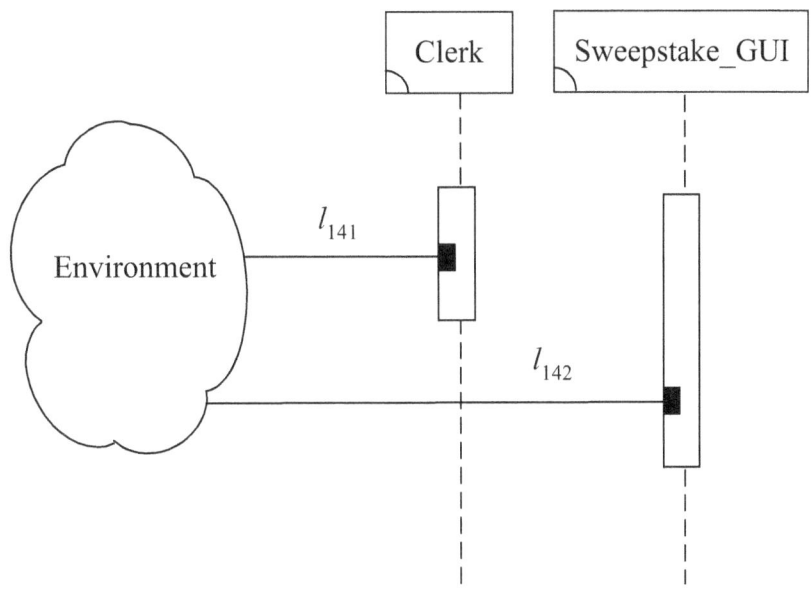

Figure 14-3   First IFD of the *Department Store's Car Sweepstakes Sales Promotion System*

The second IFD of the operation-based infinite-queue SBC process of the *Department Store's Car Sweepstakes Sales Promotion System* is shown in Figure 14-4.

Figure 14-4   Second IFD of the *Department Store's Car Sweepstakes Sales Promotion System*

The third IFD of the operation-based infinite-queue SBC process of the *Department Store's Car Sweepstakes Sales Promotion System* is shown in Figure 14-5.

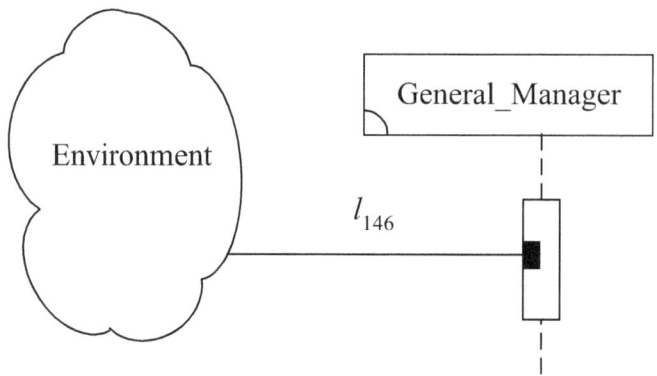

Figure 14-5   Third IFD of the *Department Store's Car Sweepstakes Sales Promotion System*

14-2 Prefixes of the Department Store's Car Sweepstakes Sales Promotion System

The t_{141} (operation-based called action with no conditions and no variable settings) prefix is defined as "<OPERATION_CALL, *Clerk*, CALLED, l_{141}>", as shown in Figure 14-6.

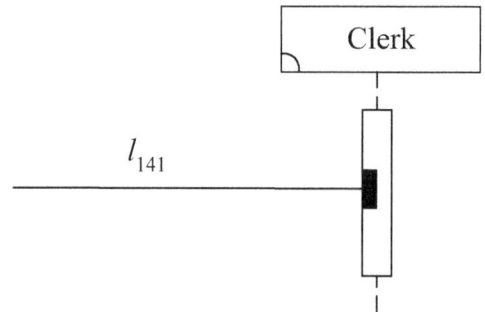

Figure 14-6   Prefix of t_{141}

The t_{142} (operation-based called action with no conditions and no variable settings) prefix is defined as "<OPERATION_CALL, *Sweepstake_GUI*, CALLED, l_{142}>", as shown in Figure 14-7.

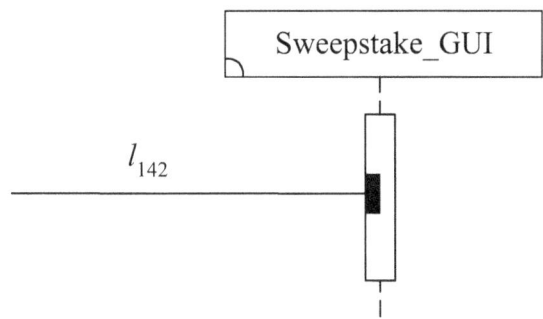

Figure 14-7   Prefix of t_{142}

The t_{143} (operation-based called action with no conditions and no variable settings) prefix is defined as "<OPERATION_CALL, *Sweepstake_Host*, CALLED, l_{143}>", as shown in Figure 14-8.

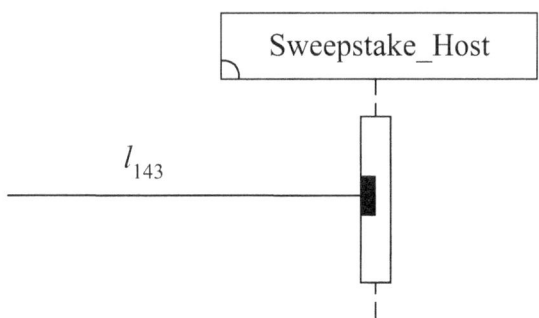

Figure 14-8   Prefix of t_{143}

The t_{144} (operation-based interaction with no conditions and no variable settings) prefix is defined as "<OPERATION_CALL, *Sweepstake_Host*, l_{144}, *Sweepstake_GUI*>", as shown in Figure 14-9.

130

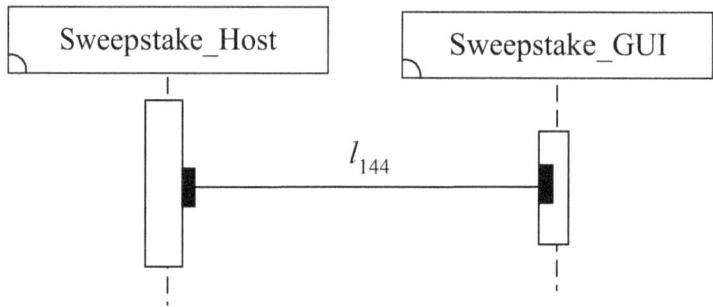

Figure 14-9   Prefix of t_{144}

The t_{145} (operation-based called action with no conditions and no variable settings) prefix is defined as "<OPERATION_RETURN, *Sweepstake_Host*, CALLED, l_{145}>", as shown in Figure 14-10.

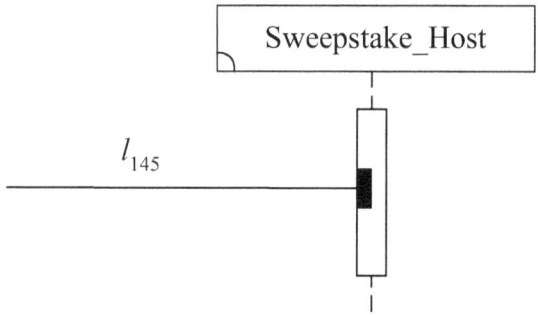

Figure 14-10   Prefix of t_{145}

The t_{146} (operation-based called action with no conditions and no variable settings) prefix is defined as "<OPERATION_CALL, *General_Manager*, CALLED, l_{146}>", as shown in Figure 14-11.

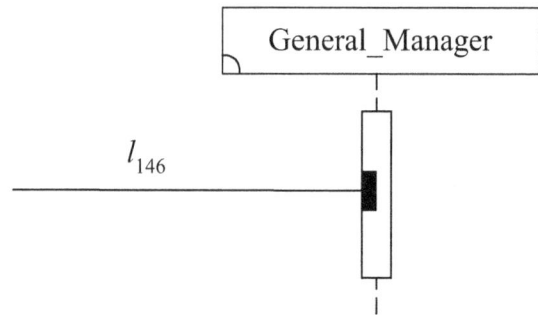

Figure 14-11   Prefix of t_{146}

Figure 14-12 shows all operation call or operation return formulas of the operation-based infinite-queue SBC process of the *Department Store's Car Sweepstakes Sales Promotion System.*

Entity name	Operation Call Formula	Operation Return Formula
l_{141}	Buy_More_Than_Thousand_Dollars(Out Sweepstake_Number)	
l_{142}	Register(In Sweepstake_Number, Personal_Data)	
l_{143}	Draw_Out	
l_{144}	Lucky_Draw(Out Winners_List)	
l_{145}		Draw_Out(Out Winners_List)
l_{146}	Award(In Personal_Data; Out Car)	

Figure 14-12 Operation Call or Operation Return Formulas of the *Department Store's Car Sweepstakes Sales Promotion System*'s Process

Figure 14-13 shows the primitive data type specification of the *Sweepstake_Number* and *Car* parameters.

Parameter	Data Type	Instances
Sweepstake_Number	Text	ABC12345
Car	Vehicle	Toyota Camry 2000

Figure 14-13   Primitive Data Type Specification

Figure 14-14 shows the composite data type specification of the *Personal_Data* input parameter occurring in the *Register(In Sweepstake_Number, Personal_Data)* and *Award(In Personal_Data; Out Car)* operation call formulas.

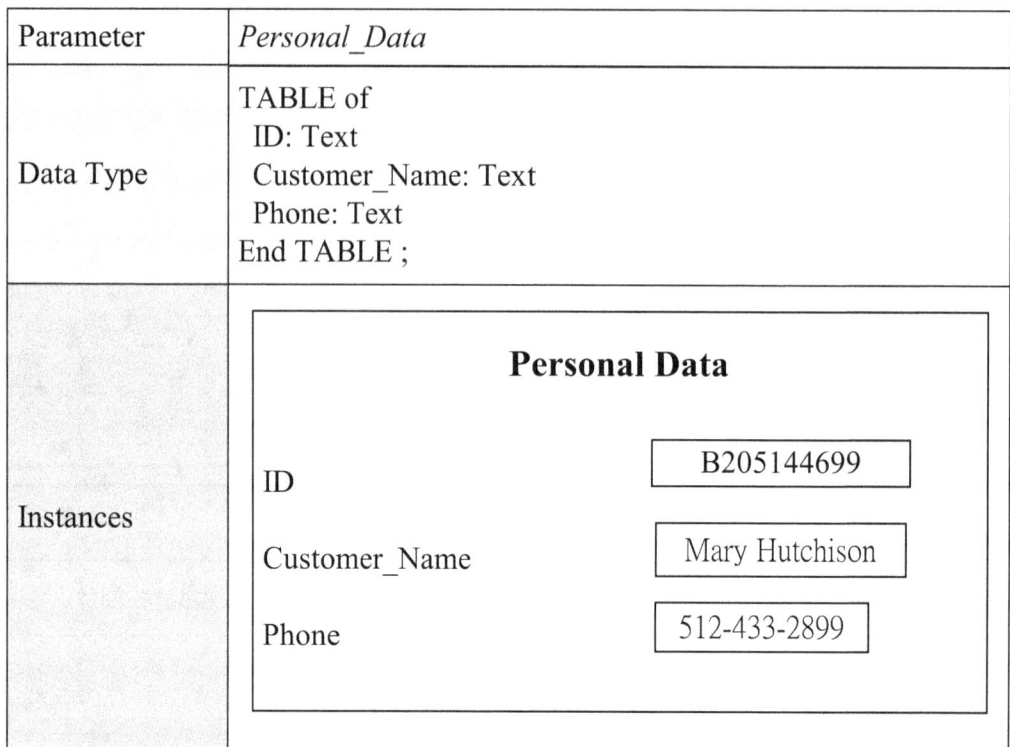

Parameter	*Personal_Data*
Data Type	TABLE of ID: Text Customer_Name: Text Phone: Text End TABLE ;
Instances	**Personal Data** ID — B205144699 Customer_Name — Mary Hutchison Phone — 512-433-2899

Figure 14-14   Composite Data Type Specification

Figure 14-15 shows the composite data type specification of the *Winners_List)* output parameter occurring in the *Lucky_Draw(Out Winners_List)* and *Draw_Out_Return(Out Winners_List)* operation call formulas.

Parameter	*Winners_List*

Data Type	TABLE of
	  ID: Text
	  Customer_Name: Text
	End TABLE ;

Instances	

ID	Customer_Name
B205144699	Mary Hutchison
P4305132187	Alice Bryant

Figure 14-15　　Composite Data Type Specification

14-3 Process of the Department Store's Car Sweepstakes Sales Promotion System

The following transition graph shows, in Figure 14-16, the semantics of P_{141}'s operation-based infinite-queue SBC process.

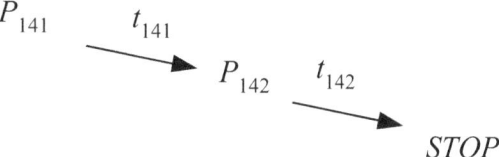

P_{141} 　 t_{141} 　 P_{142} 　 t_{142} 　 *STOP*

Figure 14-16　Transition graph of the P_{141} Process

In the transition graph of the P_{141}'s operation-based infinite-queue SBC process, processes P_{141} and P_{142} are defined as in Figure 14-17.

$$P_{141} \overset{\text{def}}{=\!=} t_{141} \bullet P_{142}$$
$$P_{142} \overset{\text{def}}{=\!=} t_{142} \bullet STOP$$

Figure 14-17　Definition of Processes P_{141} and P_{142}

The following transition graph shows, in Figure 14-18, the semantics of P_{143}'s operation-based infinite-queue SBC process.

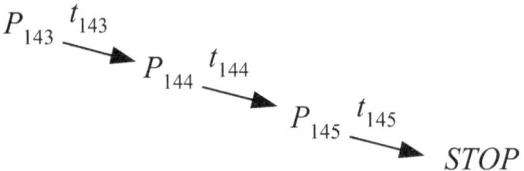

Figure 14-18   Transition graph of the P_{143} Process

In the transition graph of the P_{143}'s operation-based infinite-queue SBC process, processes P_{143}, P_{144} and P_{145} are defined as in Figure 14-19.

$$P_{143} \overset{\text{def}}{=\joinrel=} t_{143} \bullet P_{144}$$

$$P_{144} \overset{\text{def}}{=\joinrel=} t_{144} \bullet P_{145}$$

$$P_{145} \overset{\text{def}}{=\joinrel=} t_{145} \bullet STOP$$

Figure 14-19   Definition of Processes P_{143}, P_{144}, and P_{145}

The following transition graph shows, in Figure 14-20, the semantics of P_{146}'s operation-based infinite-queue SBC process.

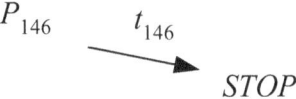

Figure 14-20   Transition graph of the P_{146} Process

In the transition graph of the P_{146}'s operation-based infinite-queue SBC process, process P_{146} is defined as in Figure 14-21.

$$P_{146} \stackrel{\text{def}}{=\!=} t_{146} \bullet STOP$$

Figure 14-21   Definition of Process P_{146}

The operation-based infinite-queue SBC process of the *Department Store's Car Sweepstakes Sales Promotion System*, A_{141}, is defined as "$!P_{141}\|!P_{143}\|!P_{146}$".

Chapter 15: Operation-Based Infinite-Queue SBC Process of the Female Mantis

In this chapter, we use the operation-based infinite-queue SBC process algebra to model the *Female Mantis* as shown in Figure 15-1.

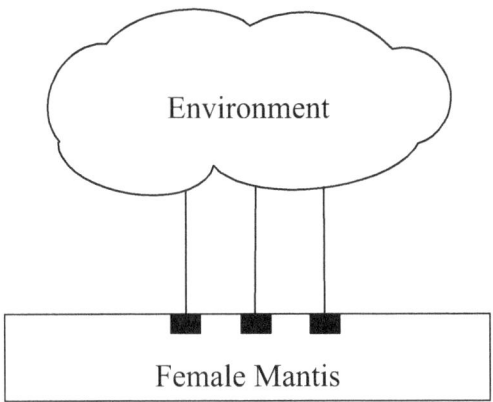

Figure 15-1   Systems Modeling the *Female Mantis*

15-1 BNF Tree of the Female Mantis

The operation-based infinite-queue SBC process of the *Female Mantis*, A_{201}, is defined as "$!(t_{201} \bullet t_{202} \bullet t_{203} \bullet STOP) \| !(t_{204} \bullet t_{205} \bullet STOP) \| !(t_{206} \bullet t_{207} \bullet STOP)$".

We draw the operation-based infinite-queue SBC process algebra Backus-Naur Form tree of A_{201} as shown in Figure 15-2.

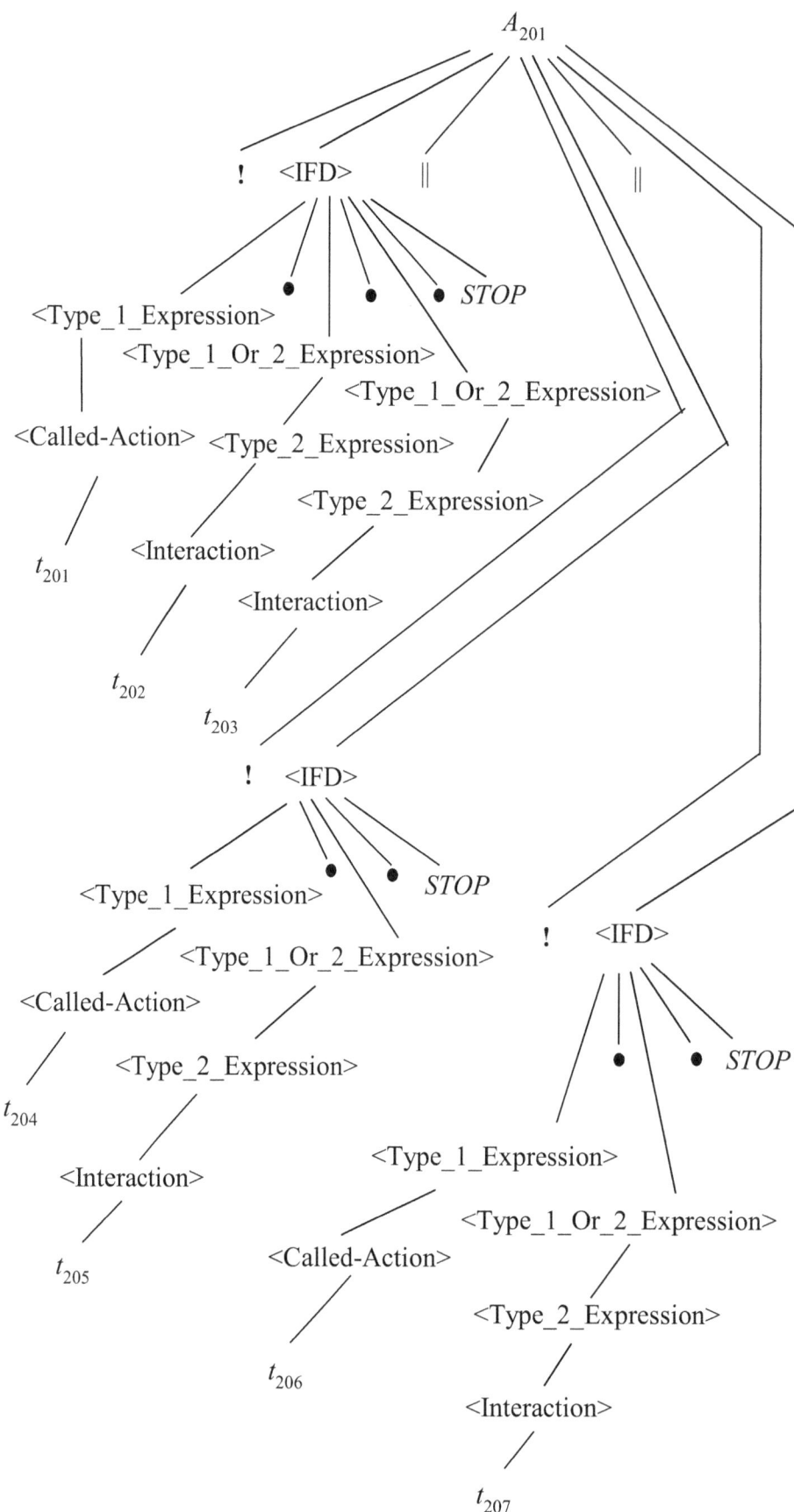

Figure 15-2   Backus-Naur Form Tree of the *Female Mantis'*
Operation-Based Infinite-Queue SBC Process

There are three IFDs in the operation-based infinite-queue SBC process of the *Female Mantis*. The first IFD is shown in Figure 15-3.

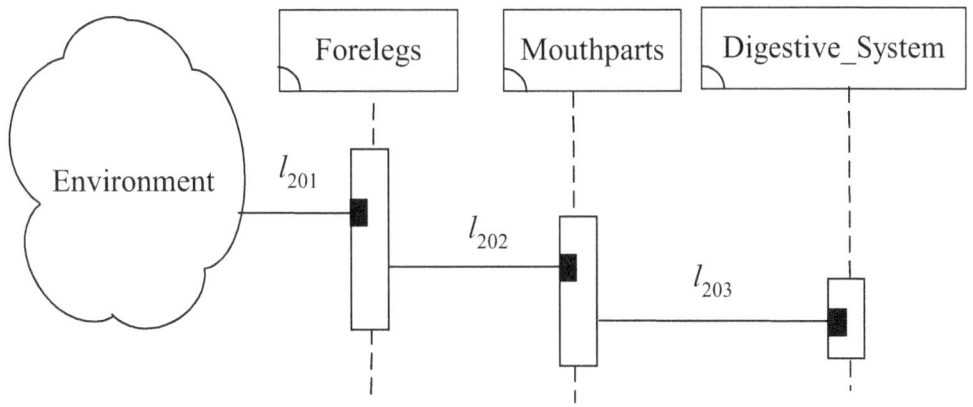

Figure 15-3   First IFD of the *Female Mantis*

The second IFD of the operation-based infinite-queue SBC process of the *Female Mantis* is shown in Figure 15-4.

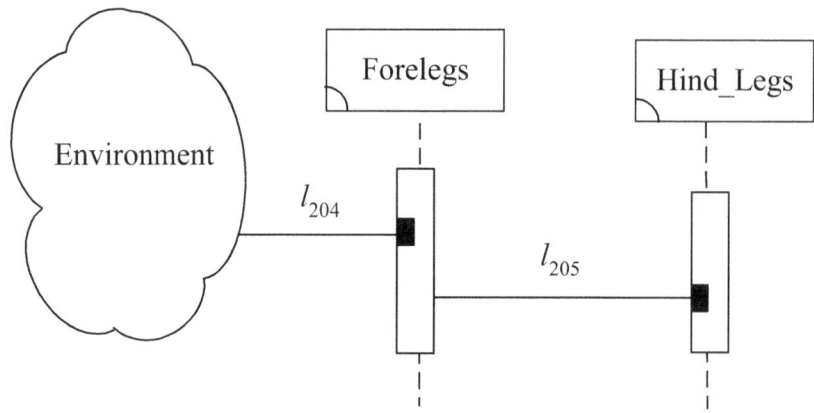

Figure 15-4   Second IFD of the *Female Mantis*

The third IFD of the operation-based infinite-queue SBC process of the *Female Mantis* is shown in Figure 15-5.

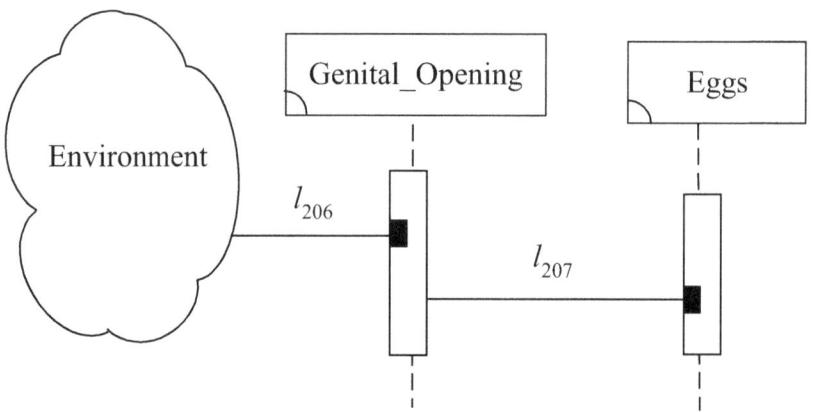

Figure 15-5   Third IFD of the *Female Mantis*

15-2 Prefixes of the Female Mantis

The t_{201} (operation-based called action with no conditions and no variable settings) prefix is defined as "<OPERATION_CALL, *Forelegs*, CALLED, l_{201}>", as shown in Figure 15-6.

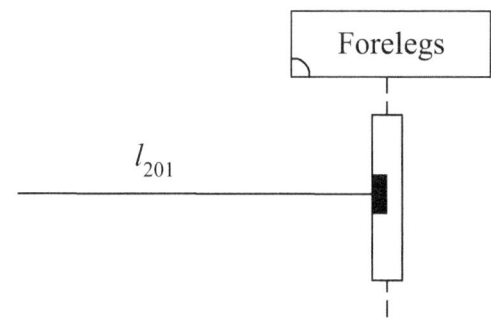

Figure 15-6   Prefix of t_{201}

The t_{202} (operation-based interaction with no conditions and no variable settings) prefix is defined as "<OPERATION_CALL, *Forelegs, l₂₀₂, Mouthparts>*", as shown in Figure 15-7.

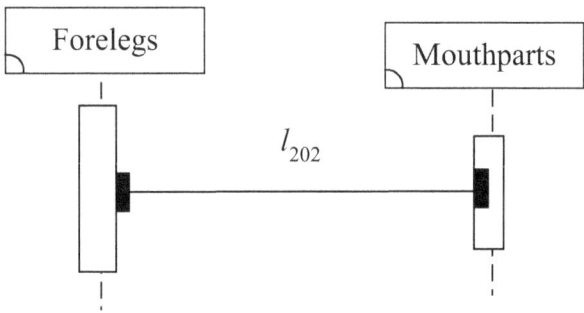

Figure 15-7   Prefix of t_{202}

The t_{203} (operation-based interaction with no conditions and no variable settings) prefix is defined as "<OPERATION_CALL, *Mouthparts, l₂₀₃, Digestive_System>*", as shown in Figure 15-8.

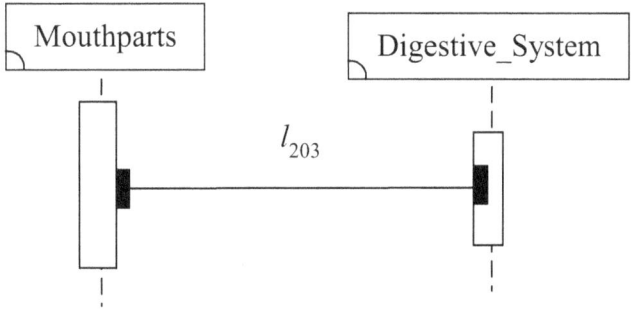

Figure 15-8   Prefix of t_{203}

The t_{204} (operation-based called action with no conditions and no variable settings) prefix is defined as "<OPERATION_CALL, *Forelegs*, CALLED, l_{204}>", as shown in Figure 15-9.

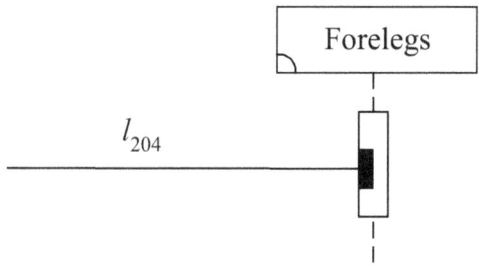

Figure 15-9   Prefix of t_{204}

The t_{205} (operation-based interaction with no conditions and no variable settings) prefix is defined as "<OPERATION_CALL, *Forelegs*, l_{205}, *Hind_Legs*>", as shown in Figure 15-10.

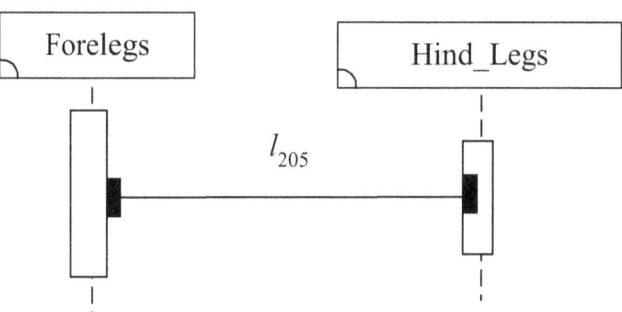

Figure 15-10   Prefix of t_{205}

The t_{206} (operation-based called action with no conditions and no variable settings) prefix is defined as "<OPERATION_CALL, *Genital_Opening*, CALLED, l_{206}>", as shown in Figure 15-11.

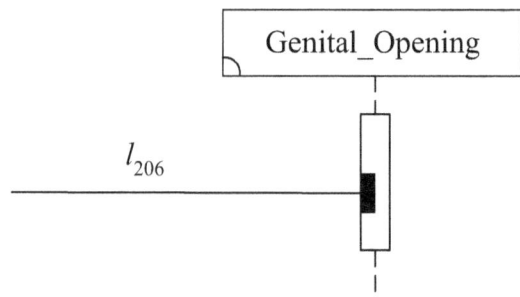

Figure 15-11   Prefix of t_{206}

The t_{207} (operation-based interaction with no conditions and no variable settings) prefix is defined as "<OPERATION_CALL, *Genital_Opening*, l_{207}, *Eggs*>", as shown in Figure 15-12.

Figure 15-12 Prefix of t_{207}

Figure 15-13 shows all operation call or operation return formulas of the operation-based infinite-queue SBC process of the *Female Mantis*.

Entity name	Operation Call Formula	Operation Return Formula
l_{201}	Grip	
l_{202}	Chew	
l_{203}	Digest	
l_{204}	Moved_forward	
l_{205}	Brought_forward	
l_{206}	Copulate	
l_{207}	Inseminate	

Figure 15-13 Operation Call or Operation Return Formulas
of the *Female Mantis'* Process

15-3 Process of the Female Mantis

The following transition graph shows, in Figure 15-14, the semantics of P_{201}'s operation-based infinite-queue SBC process.

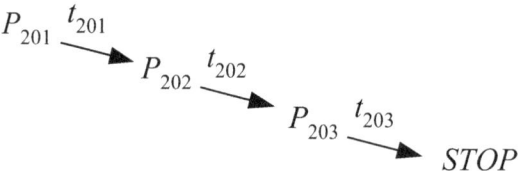

Figure 15-14   Transition graph of the P_{201} Process

In the transition graph of the P_{201}'s operation-based infinite-queue SBC process, processes P_{201}, P_{202} and P_{203} are defined as in Figure 15-15.

$$P_{201} \overset{\text{def}}{=\joinrel=} t_{201} \bullet P_{202}$$

$$P_{202} \overset{\text{def}}{=\joinrel=} t_{202} \bullet P_{203}$$

$$P_{203} \overset{\text{def}}{=\joinrel=} t_{203} \bullet STOP$$

Figure 15-15   Definition of Processes P_{201}, P_{202}, and P_{203}

The following transition graph shows, in Figure 15-16, the semantics of P_{204}'s operation-based infinite-queue SBC process.

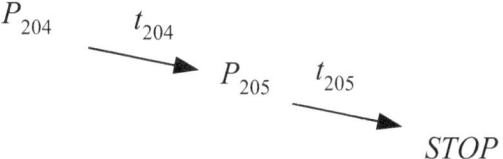

Figure 15-16   Transition graph of the P_{204} Process

In the transition graph of the P_{204}'s operation-based infinite-queue SBC process, processes P_{204} and P_{205} are defined as in Figure 15-17.

$$P_{204} \stackrel{\text{def}}{=\joinrel=} t_{204} \bullet P_{205}$$

$$P_{205} \stackrel{\text{def}}{=\joinrel=} t_{205} \bullet STOP$$

Figure 15-17   Definition of Processes P_{204} and P_{205}

The following transition graph shows, in Figure 15-18, the semantics of P_{206}'s operation-based infinite-queue SBC process.

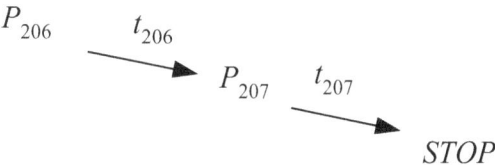

Figure 15-18   Transition graph of the P_{206} Process

In the transition graph of the P_{206}'s operation-based infinite-queue SBC process, processes P_{206} and P_{207} are defined as in Figure 15-19.

$$P_{206} \overset{\text{def}}{=\!=} t_{206} \bullet P_{207}$$

$$P_{207} \overset{\text{def}}{=\!=} t_{207} \bullet STOP$$

Figure 15-19 Definition of Processes P_{206} and P_{207}

The operation-based infinite-queue SBC process of the *Female Mantis*, A_{201}, is defined as "$!P_{201}\|!P_{204}\|!P_{206}$".

Chapter 16: Operation-Based Infinite-Queue SBC Process of the Human Body

In this chapter, we use the operation-based infinite-queue SBC process algebra to model the *Human Body* as shown in Figure 16-1.

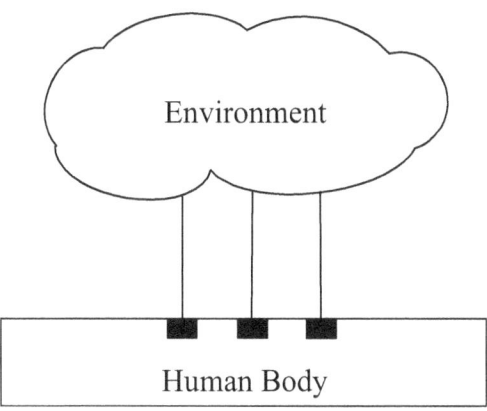

Figure 16-1   Systems Modeling the *Human Body*

16-1 BNF Tree of the Human Body

The operation-based infinite-queue SBC process of the *Human Body*, A_{251}, is defined   as "$!(t_{251} \bullet t_{252} \bullet t_{253} \bullet t_{254} \bullet t_{255} \bullet STOP) \| !(t_{256} \bullet t_{257} \bullet t_{258} \bullet t_{259} \bullet STOP) \| !(t_{260} \bullet t_{261} \bullet STOP)$".

We draw the operation-based infinite-queue SBC process algebra Backus-Naur Form tree of A_{251} as shown in Figure 16-2.

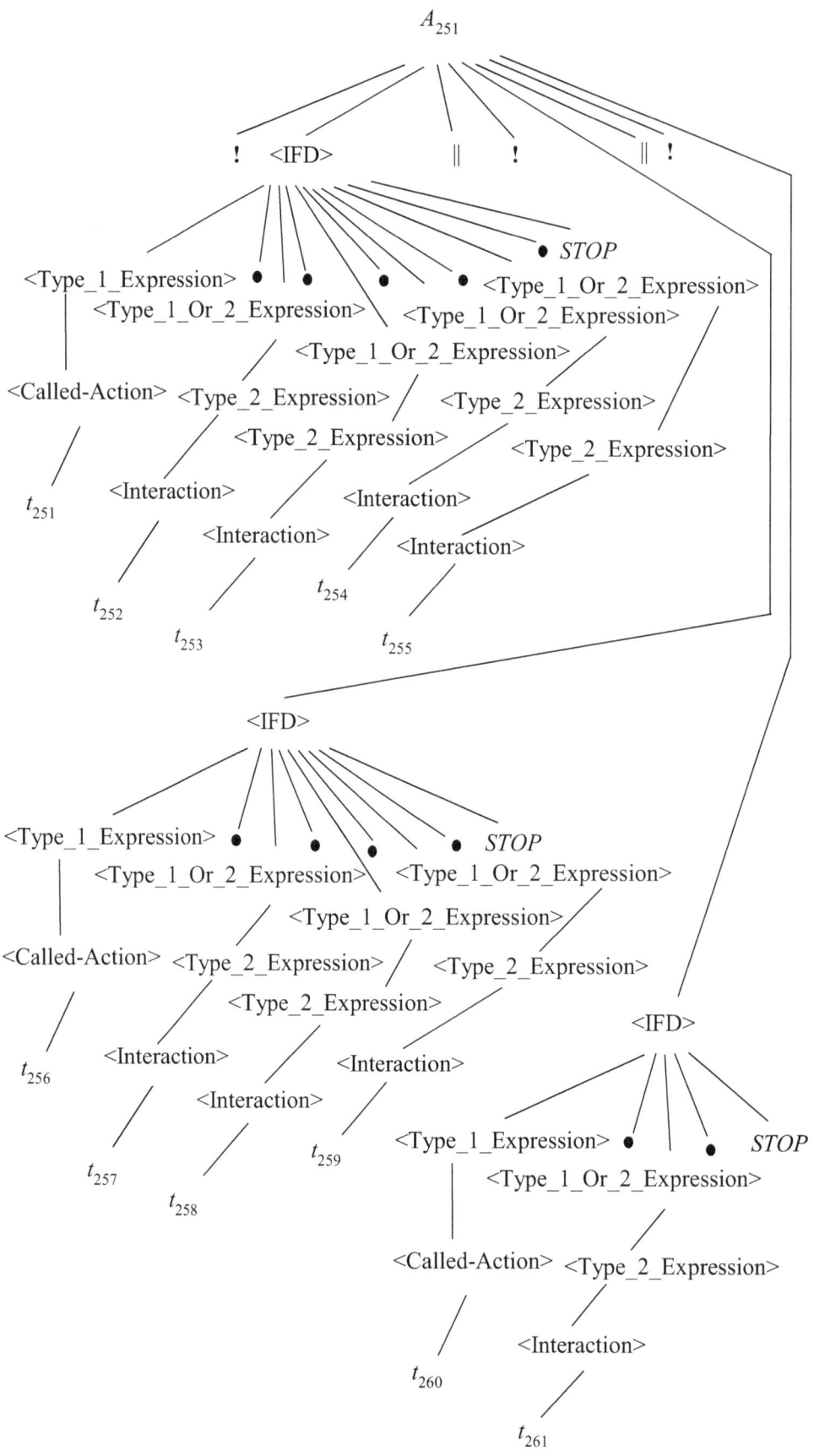

Figure 16-2   Backus-Naur Form Tree of the *Human Body*'s
Operation-Based Infinite-Queue SBC Process

There are three IFDs in the operation-based infinite-queue SBC process of the *Human Body*. The first IFD is shown in Figure 16-3.

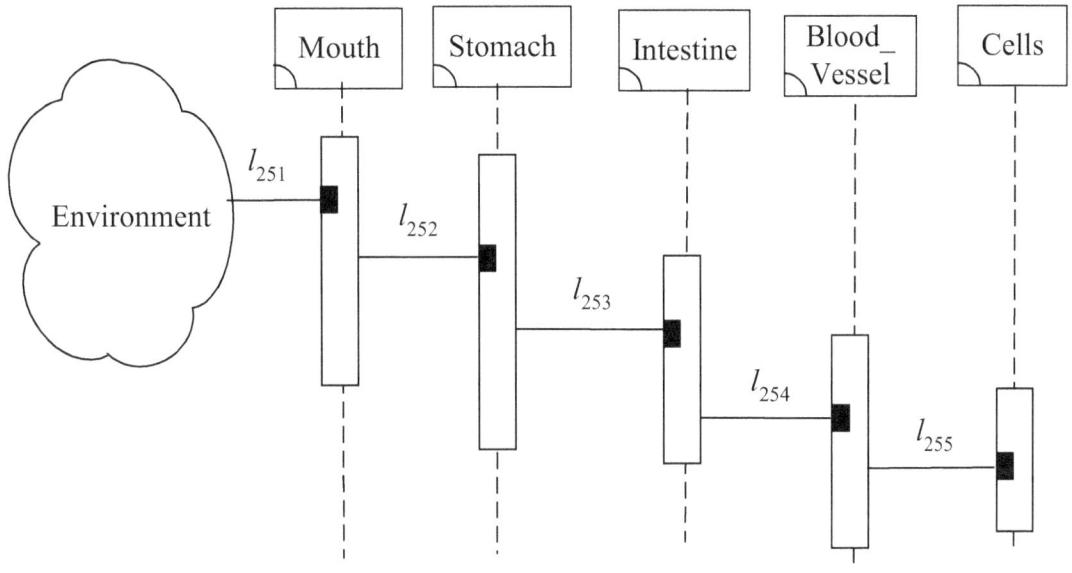

Figure 16-3   First IFD of the *Human Body*

The second IFD of the operation-based infinite-queue SBC process of the *Human Body* is shown in Figure 16-4.

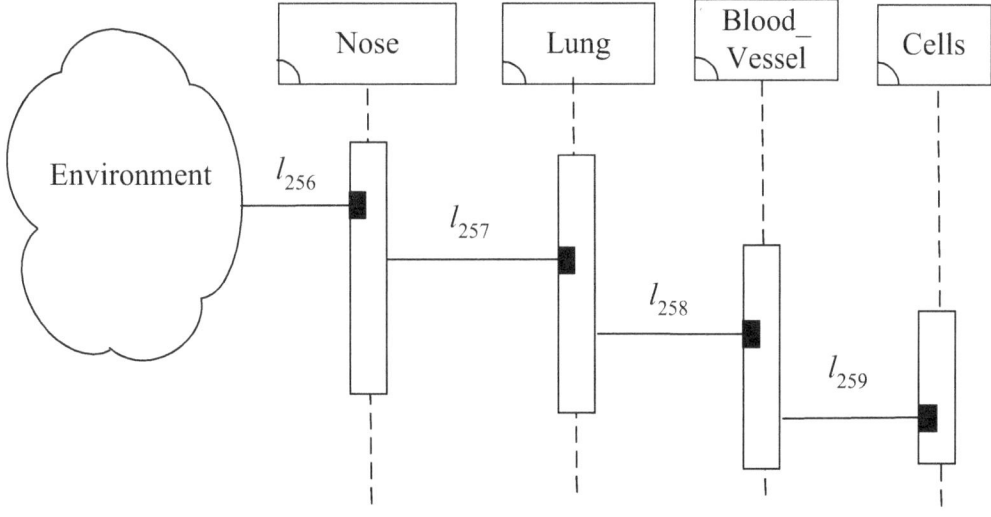

Figure 16-4   Second IFD of the *Human Body*

The third IFD of the operation-based infinite-queue SBC process of the

Human Body is shown in Figure 16-5.

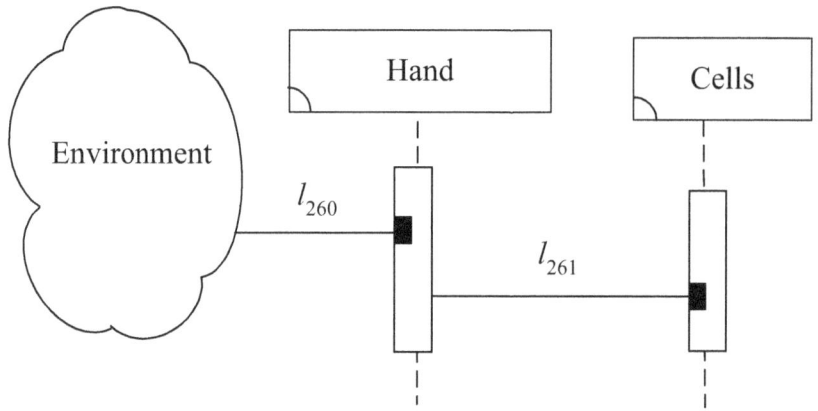

Figure 16-5   Third IFD of the *Human Body*

16-2 Prefixes of the Human Body

The t_{251} (operation-based called action with no conditions and no variable settings) prefix is defined as "<OPERATION_CALL, *Mouth*, CALLED, l_{251}>", as shown in Figure 16-6.

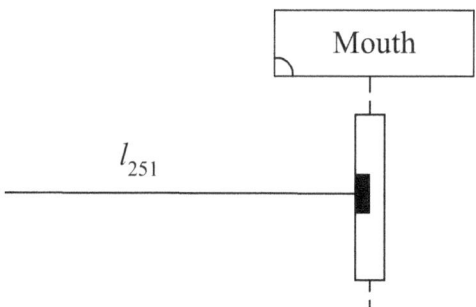

Figure 16-6   Prefix of t_{251}

The t_{252} (operation-based interaction with no conditions and no variable settings) prefix is defined as "<OPERATION_CALL, *Mouth*, l_{252}, *Stomach*>", as shown in Figure 16-7.

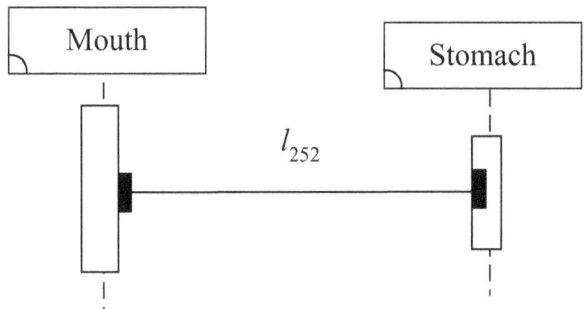

Figure 16-7   Prefix of t_{252}

The t_{253} (operation-based interaction with no conditions and no variable settings) prefix is defined as "<OPERATION_CALL, *Stomach*, l_{253}, *Intestine*>", as shown in Figure 16-8.

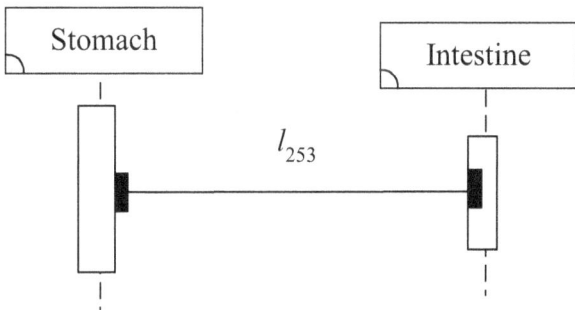

Figure 16-8   Prefix of t_{253}

The t_{254} (operation-based interaction with no conditions and no variable settings) prefix is defined as "<OPERATION_CALL, *Intestine*, l_{254}, *Blood_Vessel*>", as shown in Figure 16-9.

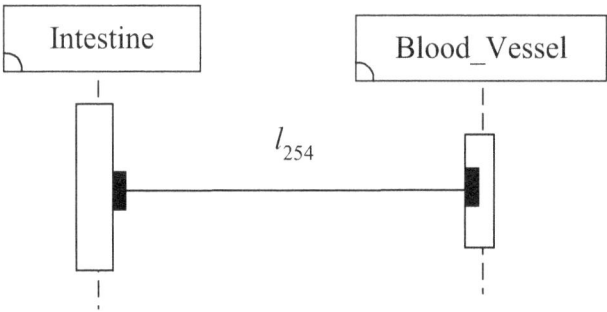

Figure 16-9   Prefix of t_{254}

The t_{255} (operation-based interaction with no conditions and no variable settings) prefix is defined as "<OPERATION_CALL, *Blood_Vessel*, l_{255}, *Cells*>", as shown in Figure 16-10.

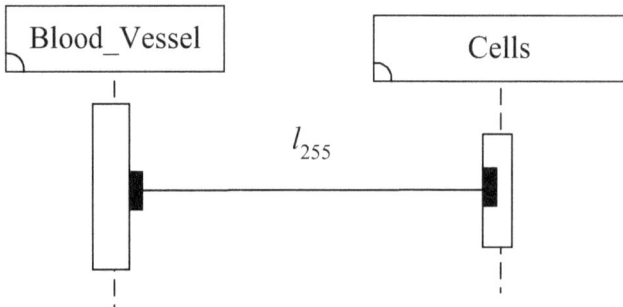

Figure 16-10　Prefix of t_{255}

The t_{256} (operation-based called action with no conditions and no variable settings) prefix is defined as "<OPERATION_CALL, *Nose*, CALLED, l_{256}>", as shown in Figure 16-11.

Figure 16-11　Prefix of t_{256}

The t_{257} (operation-based interaction with no conditions and no variable settings) prefix is defined as "<OPERATION_CALL, *Nose*, l_{257}, *Lung*>", as shown in Figure 16-12.

Figure 16-12　Prefix of t_{257}

The t_{258} (operation-based interaction with no conditions and no variable settings) prefix is defined as "<OPERATION_CALL, *Lung*, l_{258}, *Blood_Vessel*>", as shown in Figure 16-13.

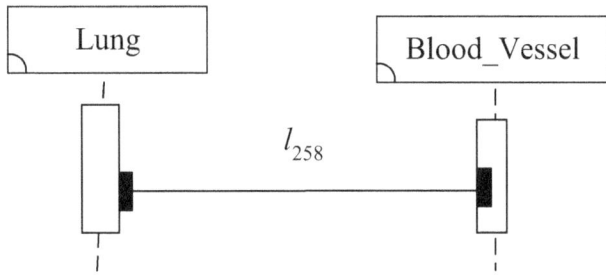

Figure 16-13   Prefix of t_{258}

The t_{259} (operation-based interaction with no conditions and no variable settings) prefix is defined as "<OPERATION_CALL, *Blood_Vessel*, l_{259}, *Cells*>", as shown in Figure 16-14.

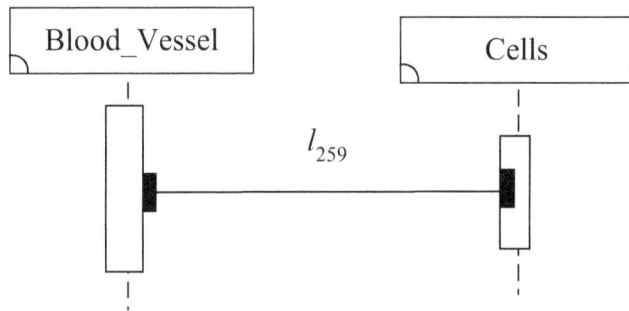

Figure 16-14   Prefix of t_{259}

The t_{260} (operation-based called action with no conditions and no variable settings) prefix is defined as "<OPERATION_CALL, *Hand*, CALLED, l_{260}>", as shown in Figure 16-15.

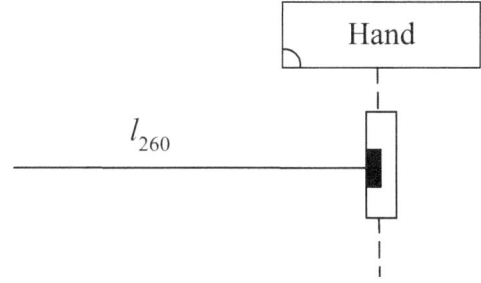

Figure 16-15   Prefix of t_{260}

The t_{261} (operation-based interaction with no conditions and no variable settings) prefix is defined as "<OPERATION_CALL, *Hand*, l_{261}, *Cells*>", as shown in Figure 16-16.

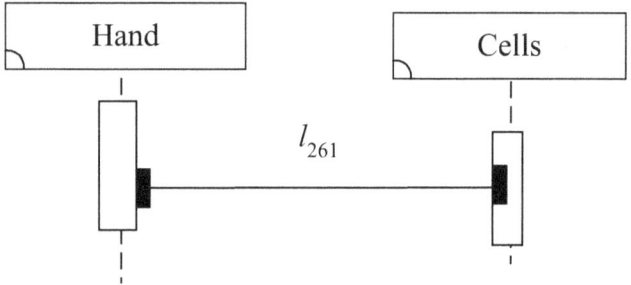

Figure 16-16   Prefix of t_{261}

Figure 16-17 shows all operation call or operation return formulas of the operation-based infinite-queue SBC process of the *Human Body*.

Entityname	Operation Call Formula	Operation Return Formula
l_{251}	Chew	
l_{252}	Digest	
l_{253}	Absorb_nutrients	
l_{254}	Transport_nutrients	
l_{255}	Store_nutrients	
l_{256}	Breathe	
l_{257}	Exchange_gas	
l_{258}	Transport_gas	
l_{259}	Respire	
l_{260}	Punch	
l_{261}	Consume_nutrients	

Figure 16-17   Operation Call or Operation Return Formulas of
the *Human Body*'s Process

16-3 Process of the Human Body

The following transition graph shows, in Figure 16-18, the semantics of P_{251}'s operation-based infinite-queue SBC process.

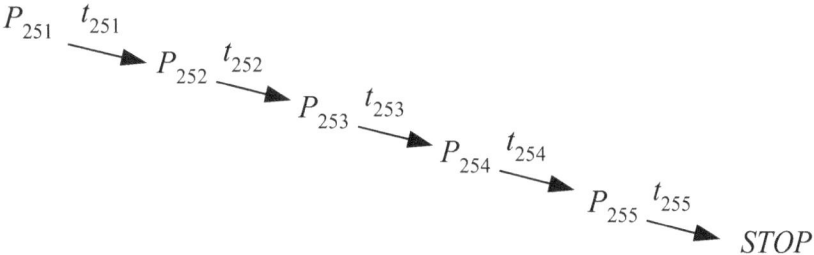

Figure 16-18   Transition graph of the P_{251} Process

In the transition graph of the P_{251}'s operation-based infinite-queue SBC process, processes P_{251}, P_{252}, P_{253}, P_{254} and P_{255} are defined as in Figure 16-19.

$$P_{251} \overset{\text{def}}{=\!=} t_{251} \bullet P_{252}$$

$$P_{252} \overset{\text{def}}{=\!=} t_{252} \bullet P_{253}$$

$$P_{253} \overset{\text{def}}{=\!=} t_{253} \bullet P_{254}$$

$$P_{254} \overset{\text{def}}{=\!=} t_{254} \bullet P_{255}$$

$$P_{255} \overset{\text{def}}{=\!=} t_{255} \bullet STOP$$

Figure 16-19   Definition of Processes P_{251}, P_{252}, P_{253}, P_{254}, and P_{255}

The following transition graph shows, in Figure 16-20, the semantics of P_{256}'s operation-based infinite-queue SBC process.

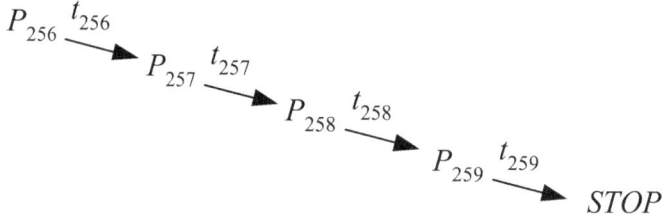

Figure 16-20   Transition graph of the P_{256} Process

In the transition graph of the P_{256}'s operation-based infinite-queue SBC process, processes P_{256}, P_{257}, P_{258} and P_{259} are defined as in Figure 16-21.

$$P_{256} \stackrel{\mathrm{def}}{=\joinrel=} t_{256} \bullet P_{257}$$

$$P_{257} \stackrel{\mathrm{def}}{=\joinrel=} t_{257} \bullet P_{258}$$

$$P_{258} \stackrel{\mathrm{def}}{=\joinrel=} t_{258} \bullet P_{259}$$

$$P_{259} \stackrel{\mathrm{def}}{=\joinrel=} t_{259} \bullet STOP$$

Figure 16-21   Definition of Processes P_{256}, P_{257}, P_{258}, and P_{259}

The following transition graph shows, in Figure 16-22, the semantics of P_{260}'s operation-based infinite-queue SBC process.

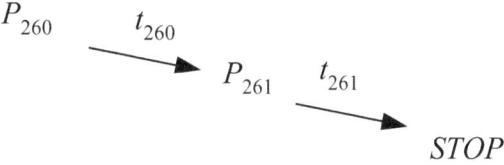

Figure 16-22   Transition graph of the P_{260} Process

In the transition graph of the P_{260}'s operation-based infinite-queue SBC process, processes P_{260} and P_{261} are defined as in Figure 16-23.

$$P_{260} \stackrel{\text{def}}{=\!=} t_{260} \bullet P_{261}$$

$$P_{261} \stackrel{\text{def}}{=\!=} t_{261} \bullet STOP$$

Figure 16-23   Definition of Processes P_{260} and P_{261}

The operation-based infinite-queue SBC process of the *Human Body*, A_{251}, is defined as "$!P_{251}\|!P_{256}\|!P_{260}$".

Chapter 17: Operation-Based Infinite-Queue SBC Process of the Disaster

In this chapter, we use the operation-based infinite-queue SBC process algebra to model the *Disaster* as shown in Figure 17-1.

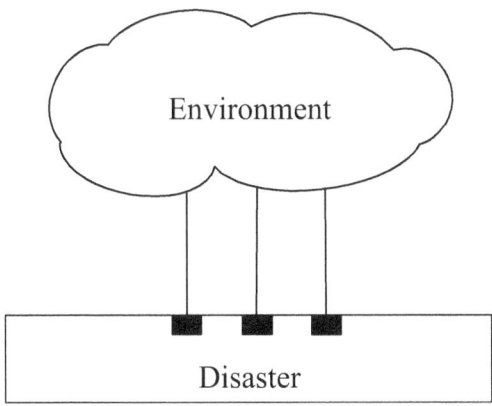

Figure 17-1   Systems Modeling the *Disaster*

17-1 BNF Tree of the Disaster

The operation-based infinite-queue SBC process of the *Disaster*, A_{301}, is defined   as "$!(t_{301} \bullet t_{302} \bullet t_{303} \bullet t_{304} \bullet STOP) \| !(t_{305} \bullet t_{306} \bullet STOP) \| !(t_{307} \bullet t_{308} \bullet t_{309} \bullet t_{310} \bullet STOP)$".

We draw the operation-based infinite-queue SBC process algebra Backus-Naur Form tree of A_{301} as shown in Figure 17-2.

160

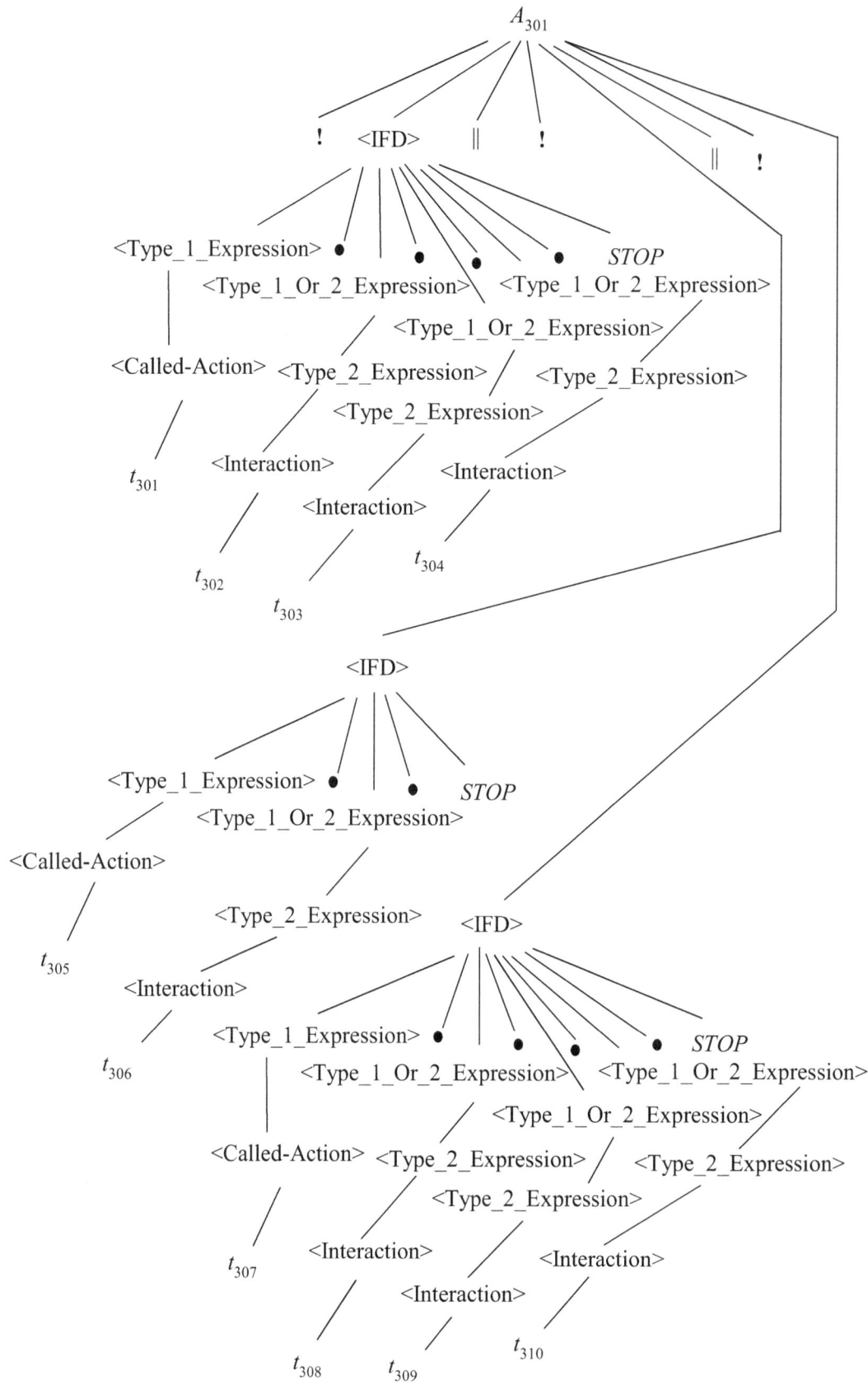

Figure 17-2   Backus-Naur Form Tree of the *Disaster*'s
Operation-Based Infinite-Queue SBC Process

There are three IFDs in the operation-based infinite-queue SBC process of the *Disaster*. The first IFD is shown in Figure 17-3.

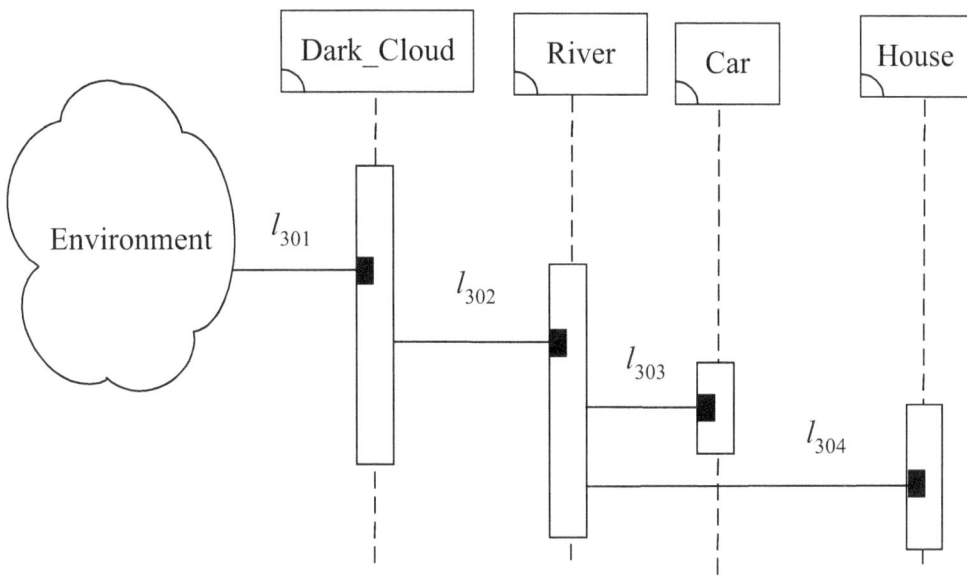

Figure 17-3   First IFD of the *Disaster*

The second IFD of the operation-based infinite-queue SBC process of the *Disaster* is shown in Figure 17-4.

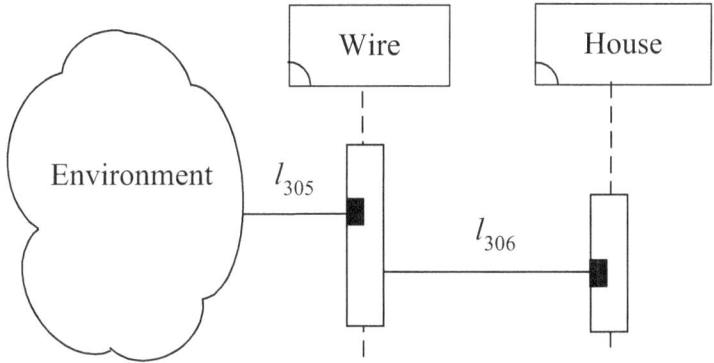

Figure 17-4   Second IFD of the *Disaster*

The third IFD of the operation-based infinite-queue SBC process of the *Disaster* is shown in Figure 17-5.

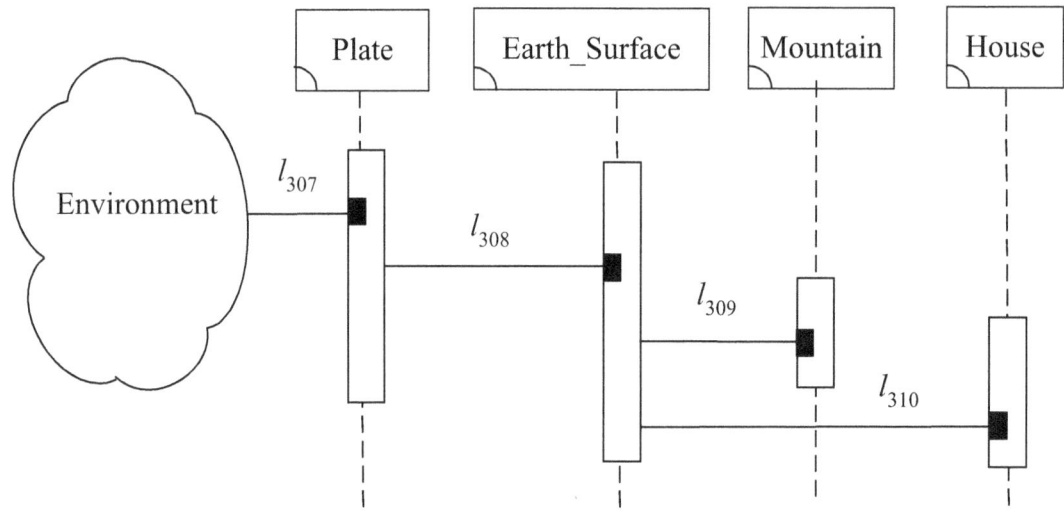

Figure 17-5   Third IFD of the *Disaster*

17-2 Prefixes of the Disaster

The t_{301} (operation-based called action with no conditions and no variable settings) prefix is defined as "<OPERATION_CALL, *Dark_Cloud*, CALLED, l_{301}>", as shown in Figure 17-6.

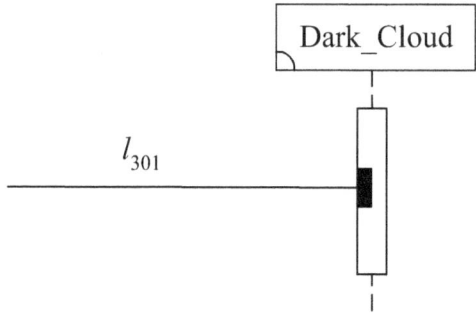

Figure 17-6   Prefix of t_{301}

The t_{302} (operation-based interaction with no conditions and no variable settings) prefix is defined as "<OPERATION_CALL, *Dark_Cloud*, l_{302}, *River*>", as shown in Figure 17-7.

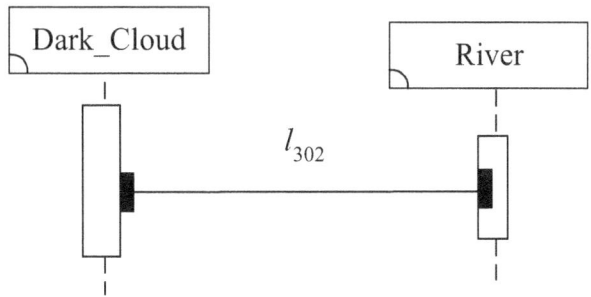

Figure 17-7   Prefix of t_{302}

The t_{303} (operation-based interaction with no conditions and no variable settings) prefix is defined as "<OPERATION_CALL, *River*, l_{303}, *Car*>", as shown in Figure 17-8.

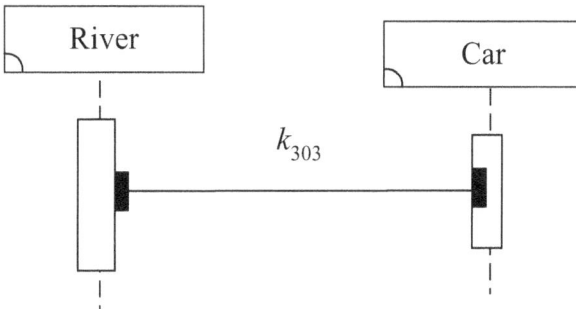

Figure 17-8   Prefix of t_{303}

The t_{304} (operation-based interaction with no conditions and no variable settings) prefix is defined as "<OPERATION_CALL, *River*, l_{304}, *House*>", as shown in Figure 17-9.

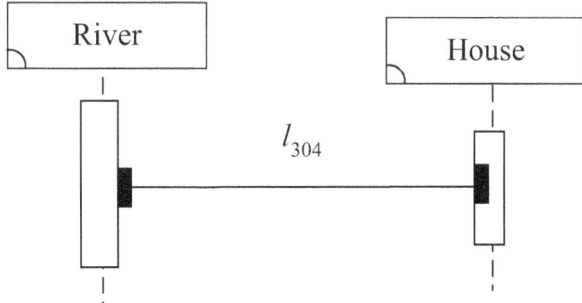

Figure 17-9   Prefix of t_{304}

The t_{305} (operation-based called action with no conditions and no variable settings) prefix is defined as "<OPERATION_CALL, *Wire*, CALLED, l_{305}>", as shown in Figure 17-10.

Figure 17-10   Prefix of t_{305}

The t_{306} (operation-based interaction with no conditions and no variable settings) prefix is defined as "<OPERATION_CALL, *Wire*, l_{306}, *House*>", as shown in Figure 17-11.

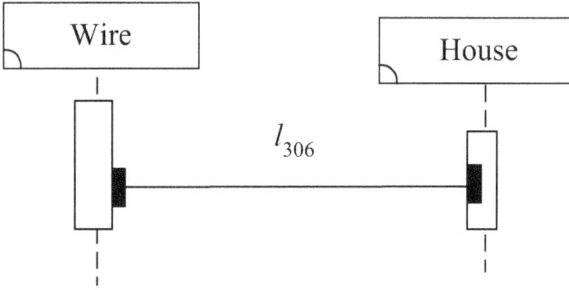

Figure 17-11   Prefix of t_{306}

The t_{307} (operation-based called action with no conditions and no variable settings) prefix is defined as "<OPERATION_CALL, *Plate*, CALLED, l_{307}>", as shown in Figure 17-12.

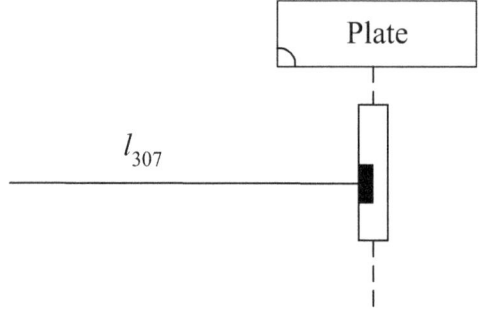

Figure 17-12   Prefix of t_{307}

The t_{308} (operation-based interaction with no conditions and no variable settings) prefix is defined as "<OPERATION_CALL, *Plate*, l_{308}, *Earth_Surface*>", as shown in Figure 17-13.

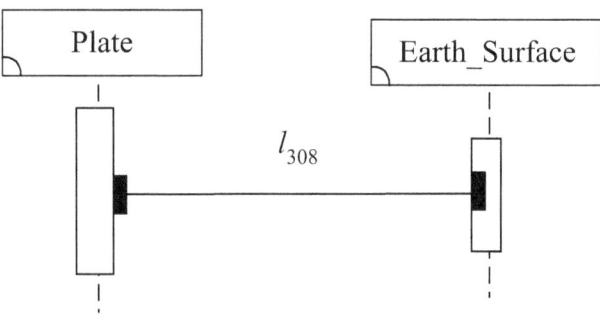

Figure 17-13   Prefix of t_{308}

The t_{309} (operation-based interaction with no conditions and no variable settings) prefix is defined as "<OPERATION_CALL, *Earth_Surface*, l_{309}, *Mountain*>", as shown in Figure 17-14.

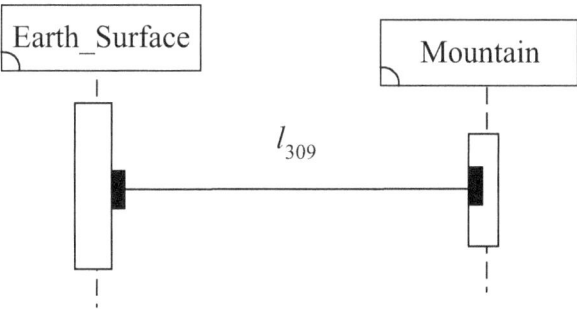

Figure 17-14   Prefix of t_{309}

166

The t_{310} (operation-based interaction with no conditions and no variable settings) prefix is defined as "<OPERATION_CALL, *Earth_Surface*, l_{310}, *House*>", as shown in Figure 17-15.

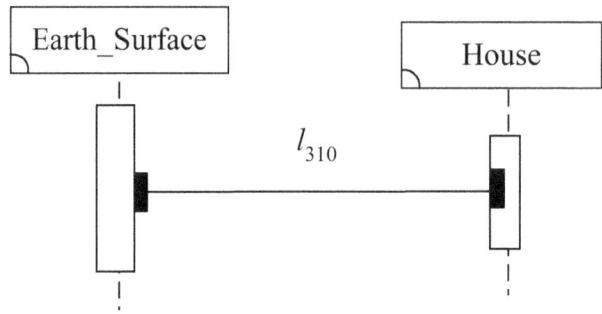

Figure 17-15   Prefix of t_{310}

Figure 17-16 shows all operation call or operation return formulas of the operation-based infinite-queue SBC process of the *Disaster*.

Entityname	Operation Call Formula	Operation Return Formula
l_{301}	Heavy_rain	
l_{302}	Overflow	
l_{303}	Soak	
l_{304}	Wash_away	
l_{305}	Bite_through	
l_{306}	Burn_down	
l_{307}	Move	
l_{308}	Ground_rupture	
l_{309}	Landslide	
l_{310}	Tilt	

Figure 17-16   Operation Call or Operation Return Formulas of
the *Disaster*'s Process

17-3 Process of the Disaster

The following transition graph shows, in Figure 17-17, the semantics of P_{301}'s operation-based infinite-queue SBC process.

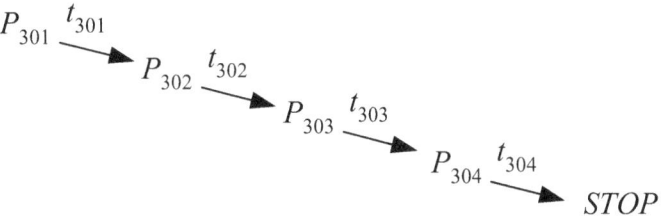

Figure 17-17   Transition graph of the P_{301} Process

In the transition graph of the P_{301}'s operation-based infinite-queue SBC process, processes P_{301}, P_{302}, P_{303} and P_{304} are defined as in Figure 17-18.

$$P_{301} \overset{\text{def}}{=\!=} t_{301} \bullet P_{302}$$

$$P_{302} \overset{\text{def}}{=\!=} t_{302} \bullet P_{303}$$

$$P_{303} \overset{\text{def}}{=\!=} t_{303} \bullet P_{304}$$

$$P_{304} \overset{\text{def}}{=\!=} t_{304} \bullet STOP$$

Figure 17-18   Definition of Processes P_{301}, P_{302}, P_{303}, and P_{304}

The following transition graph shows, in Figure 17-19, the semantics of P_{305}'s operation-based infinite-queue SBC process.

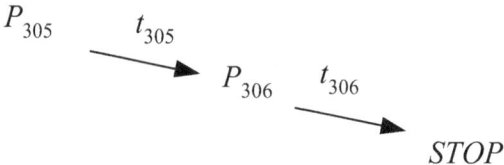

Figure 17-19   Transition graph of the P_{305} Process

In the transition graph of the P_{305}'s operation-based infinite-queue SBC process, processes P_{305} and P_{306} are defined as in Figure 17-20.

$$P_{305} \stackrel{\text{def}}{=\!=} t_{305} \bullet P_{306}$$

$$P_{306} \stackrel{\text{def}}{=\!=} t_{306} \bullet STOP$$

Figure 17-20   Definition of Processes P_{305} and P_{306}

The following transition graph shows, in Figure 17-21, the semantics of P_{307}'s operation-based infinite-queue SBC process.

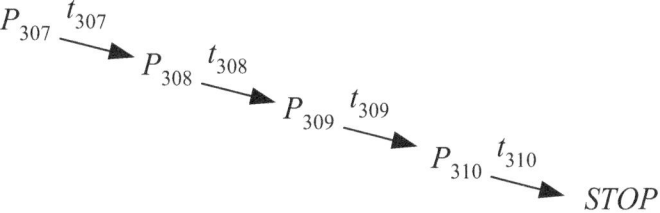

Figure 17-21   Transition graph of the P_{307} Process

In the transition graph of the P_{307}'s operation-based infinite-queue SBC process, processes P_{307}, P_{308}, P_{309} and P_{310} are defined as in Figure 17-22.

$$P_{307} \overset{\text{def}}{=\!=} t_{307} \bullet P_{308}$$

$$P_{308} \overset{\text{def}}{=\!=} t_{308} \bullet P_{309}$$

$$P_{309} \overset{\text{def}}{=\!=} t_{309} \bullet P_{310}$$

$$P_{310} \overset{\text{def}}{=\!=} t_{310} \bullet STOP$$

Figure 17-22   Definition of Processes P_{307}, P_{308}, P_{309}, and P_{310}

The operation-based infinite-queue SBC process of the *Disaster*, A_{301}, is defined as "$!P_{301}\|!P_{305}\|!P_{307}$".

Chapter 18: Operation-Based Infinite-Queue SBC Process of the Bicycle

In this chapter, we use the operation-based infinite-queue SBC process algebra to model the *Bicycle* as shown in Figure 18-1.

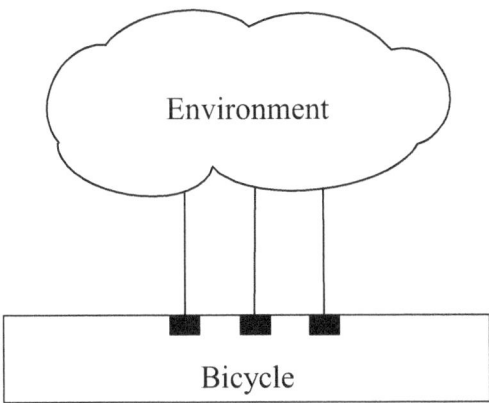

Figure 18-1　Systems Modeling the *Bicycle*

18-1 BNF Tree of the Bicycle

The operation-based infinite-queue SBC process of the *Bicycle*, A_{401}, is defined as "$!(t_{401} \bullet t_{402} \bullet t_{403} \bullet STOP)\|!(t_{404} \bullet t_{405} \bullet STOP)\|!(t_{406} \bullet t_{407} \bullet t_{408} \bullet t_{409} \bullet STOP)$".

We draw the operation-based infinite-queue SBC process algebra Backus-Naur Form tree of A_{401} as shown in Figure 18-2.

172

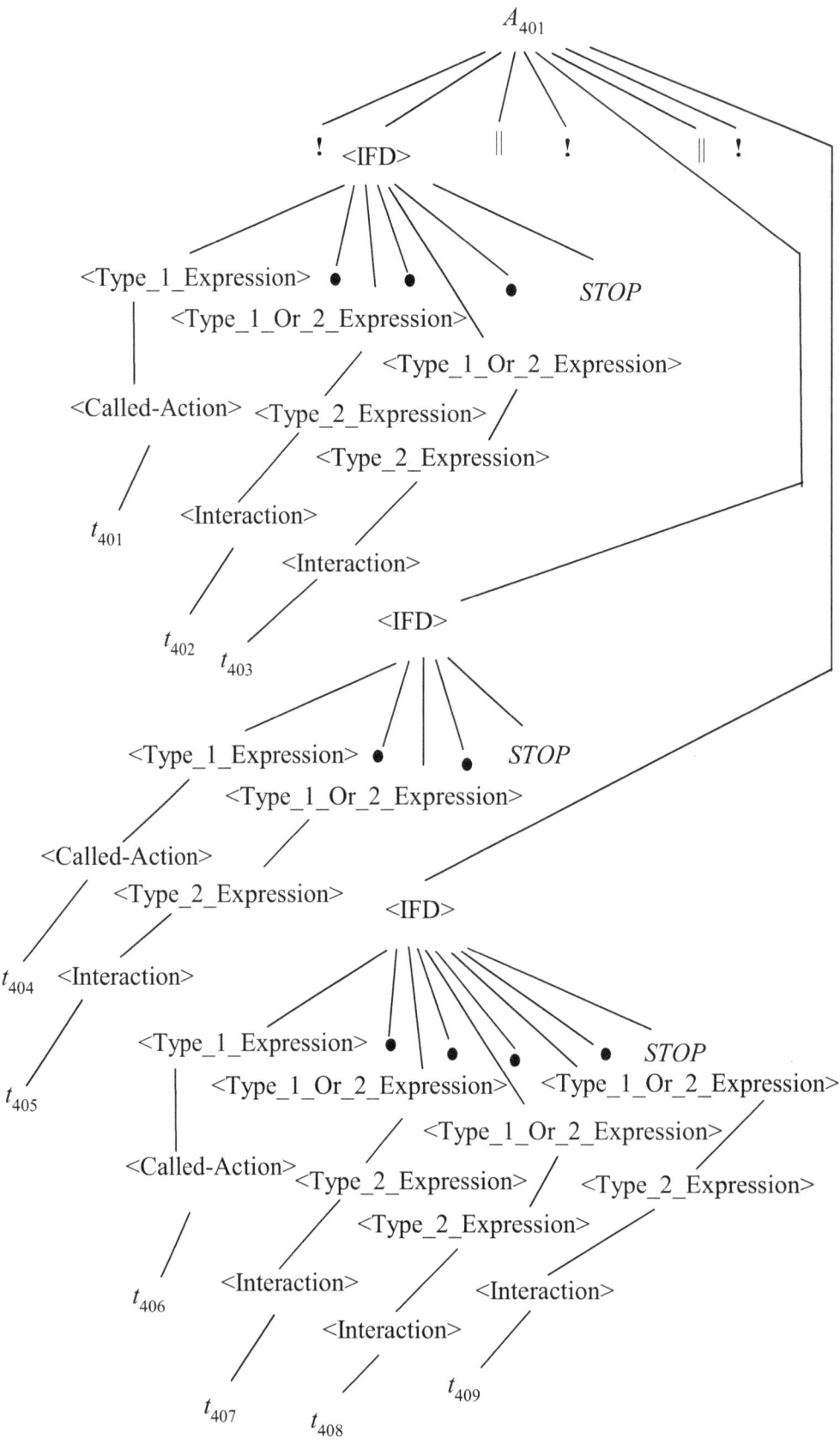

Figure 18-2   Backus-Naur Form Tree of the *Bicycle*'s
Operation-Based Infinite-Queue SBC Process

There are three IFDs in the operation-based infinite-queue SBC process of the *Bicycle*. The first IFD is shown in Figure 18-3.

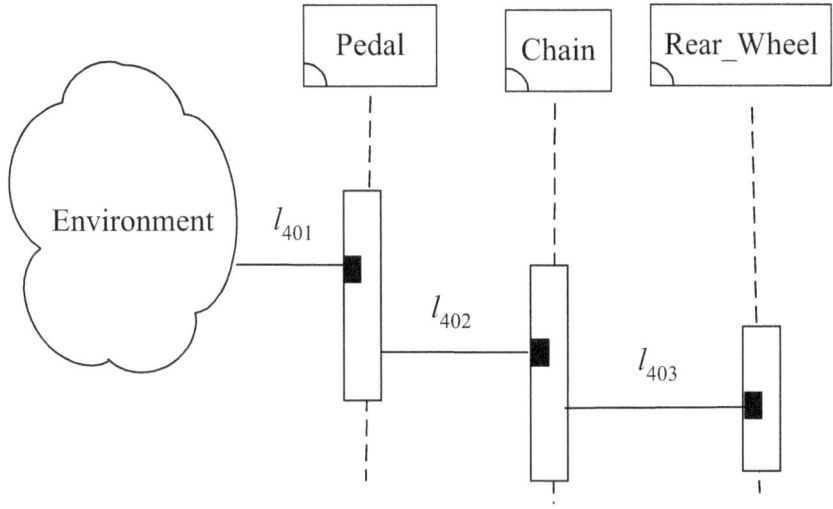

Figure 18-3   First IFD of the *Bicycle*

The second IFD of the operation-based infinite-queue SBC process of the *Bicycle* is shown in Figure 18-4.

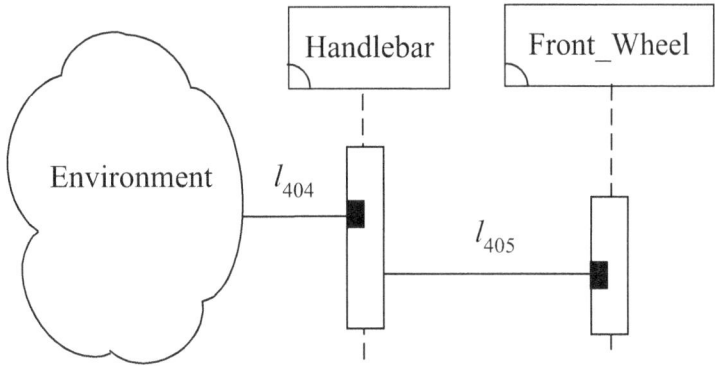

Figure 18-4   Second IFD of the *Bicycle*

The third IFD of the operation-based infinite-queue SBC process of the *Bicycle* is shown in Figure 18-5.

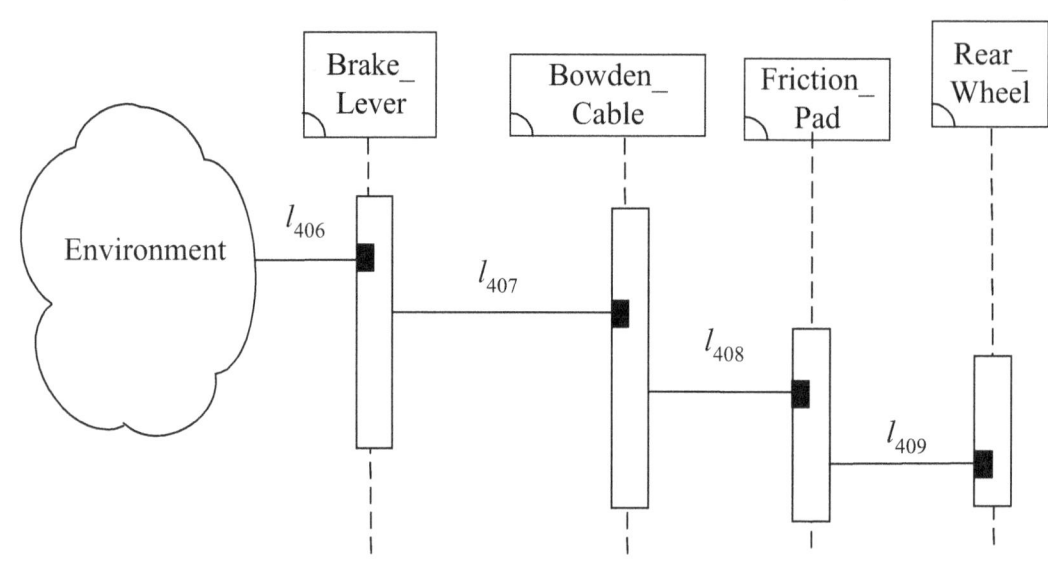

Figure 18-5   Third IFD of the *Bicycle*

18-2 Prefixes of the Bicycle

The t_{401} (operation-based called action with no conditions and no variable settings) prefix is defined as "<OPERATION_CALL, *Pedal*, CALLED, l_{401}>", as shown in Figure 18-6.

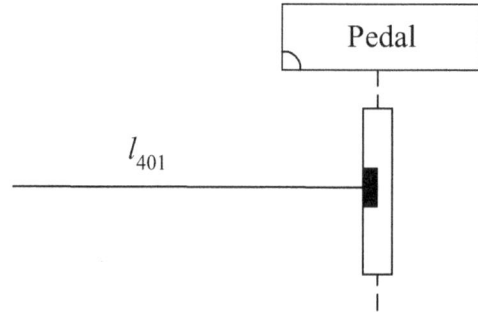

Figure 18-6   Prefix of t_{401}

The t_{402} (operation-based interaction with no conditions and no variable settings) prefix is defined as "<OPERATION_CALL, *Pedal*, l_{402}, *Chain*>", as shown in Figure 18-7.

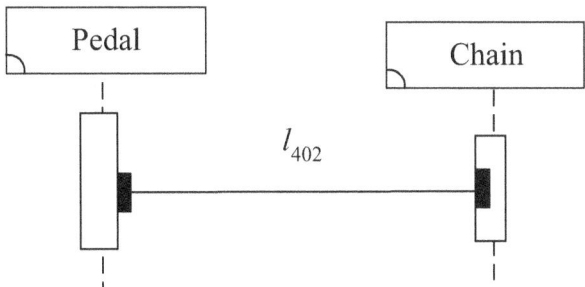

Figure 18-7   Prefix of t_{402}

The t_{403} (operation-based interaction with no conditions and no variable settings) prefix is defined as "<OPERATION_CALL, *Chain*, l_{403}, *Rear_Wheel*>", as shown in Figure 18-8.

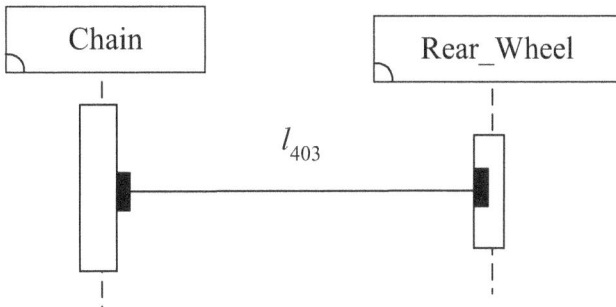

Figure 18-8   Prefix of t_{403}

The t_{404} (operation-based called action with no conditions and no variable settings) prefix is defined as "<OPERATION_CALL, *Handlebar*, CALLED, l_{404}>", as shown in Figure 18-9.

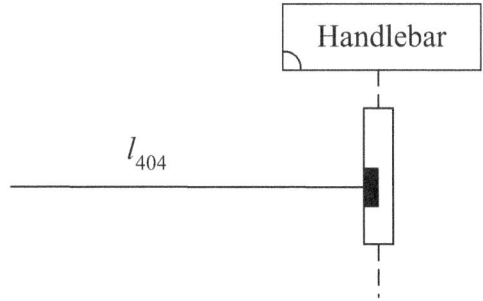

Figure 18-9   Prefix of t_{404}

The t_{405} (operation-based interaction with no conditions and no variable settings) prefix is defined as "<OPERATION_CALL, *Handlebar*, l_{405}, *Front_Wheel*>", as shown in Figure 18-10.

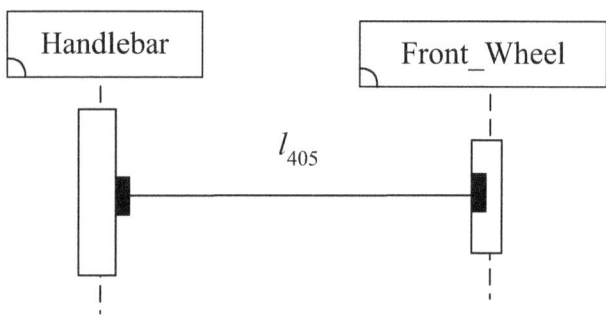

Figure 18-10   Prefix of t_{405}

The t_{406} (operation-based called action with no conditions and no variable settings) prefix is defined as "<OPERATION_CALL, *Brake_Lever*, CALLED, l_{406}>", as shown in Figure 18-11.

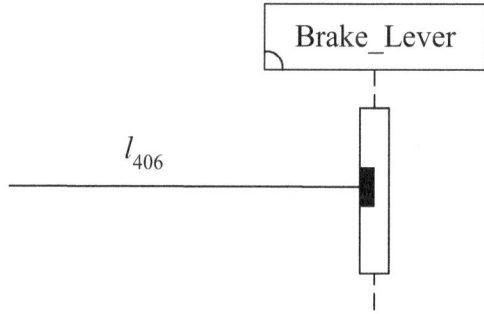

Figure 18-11   Prefix of t_{406}

The t_{407} (operation-based interaction with no conditions and no variable settings) prefix is defined as "<OPERATION_CALL, *Brake_Lever*, l_{407}, *Bowden_Cable*>", as shown in Figure 18-12.

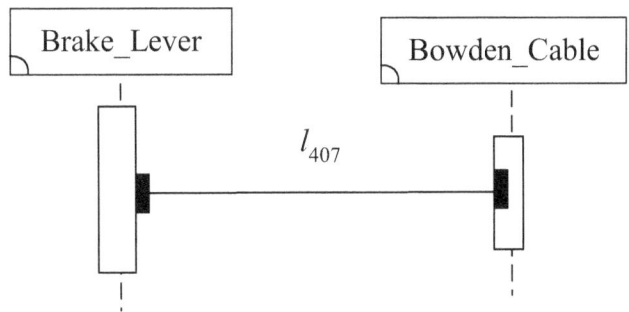

Figure 18-12   Prefix of t_{407}

The t_{408} (operation-based interaction with no conditions and no variable settings) prefix is defined as "<OPERATION_CALL, *Bowden_Cable*, l_{408}, *Friction_Pad*>", as shown in Figure 18-13.

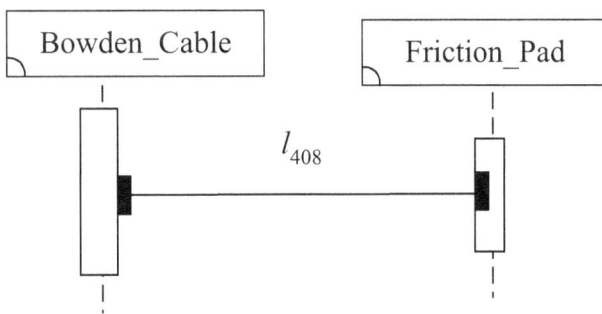

Figure 18-13   Prefix of t_{408}

The t_{409} (operation-based interaction with no conditions and no variable settings) prefix is defined as "<OPERATION_CALL, *Friction_Pad*, l_{409}, *Rear_Wheel*>", as shown in Figure 18-14.

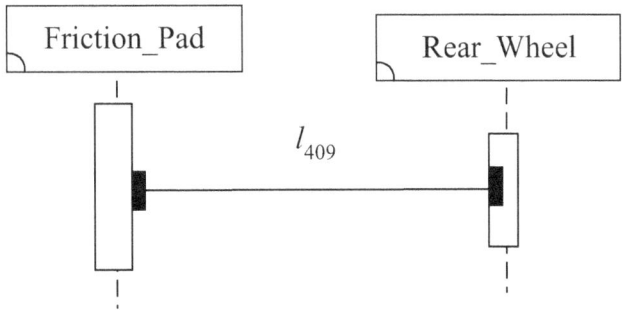

Figure 18-14   Prefix of t_{409}

Figure 18-15 shows all operation call or operation return formulas of the operation-based infinite-queue SBC process of the *Bicycle*.

Entityname	Operation Call Formula	Operation Return Formula
l_{401}	Depress	
l_{402}	Transmit_power	
l_{403}	Roll	
l_{404}	Steer	
l_{405}	Turn_left	
l_{406}	Force	
l_{407}	Pull	
l_{408}	Compress	
l_{409}	Stop_rolling	

Figure 18-15   Operation Call or Operation Return Formulas of
the *Bicycle*'s Process

18-3 Process of the Bicycle

The following transition graph shows, in Figure 18-16, the semantics of P_{401}'s operation-based infinite-queue SBC process.

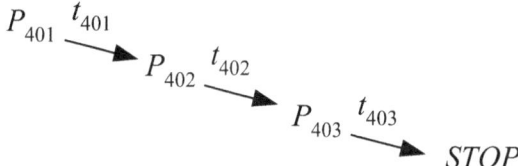

Figure 18-16   Transition graph of the P_{401} Process

In the transition graph of the P_{401}'s operation-based infinite-queue SBC process, processes P_{401}, P_{402} and P_{403} are defined as in Figure 18-17.

$$P_{401} \stackrel{\text{def}}{=\!=} t_{401} \bullet P_{402}$$

$$P_{402} \stackrel{\text{def}}{=\!=} t_{402} \bullet P_{403}$$

$$P_{403} \stackrel{\text{def}}{=\!=} t_{403} \bullet STOP$$

Figure 18-17   Definition of Processes P_{401}, P_{402}, and P_{403}

The following transition graph shows, in Figure 18-18, the semantics of P_{404}'s operation-based infinite-queue SBC process.

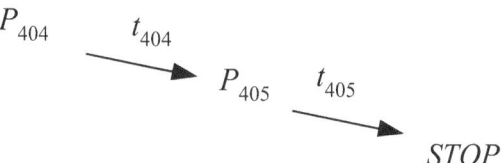

Figure 18-18   Transition graph of the P_{404} Process

In the transition graph of the P_{404}'s operation-based infinite-queue SBC process, processes P_{404} and P_{405} are defined as in Figure 18-19.

$$P_{404} \stackrel{\text{def}}{=\!=} t_{404} \bullet P_{405}$$

$$P_{405} \stackrel{\text{def}}{=\!=} t_{405} \bullet STOP$$

Figure 18-19   Definition of Processes P_{404} and P_{405}

The following transition graph shows, in Figure 18-20, the semantics of P_{406}'s operation-based infinite-queue SBC process.

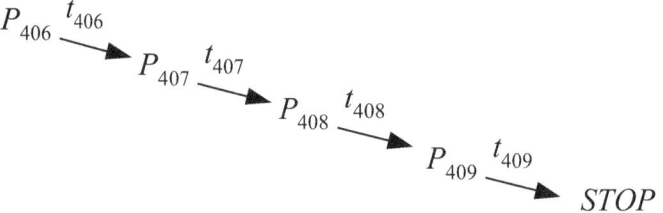

Figure 18-20   Transition graph of the P_{406} Process

In the transition graph of the P_{406}'s operation-based infinite-queue SBC process, processes P_{406}, P_{407}, P_{408} and P_{409} are defined as in Figure 18-21.

$$P_{406} \stackrel{\text{def}}{=\joinrel=} t_{406} \bullet P_{407}$$

$$P_{407} \stackrel{\text{def}}{=\joinrel=} t_{407} \bullet P_{408}$$

$$P_{408} \stackrel{\text{def}}{=\joinrel=} t_{408} \bullet P_{409}$$

$$P_{409} \stackrel{\text{def}}{=\joinrel=} t_{409} \bullet STOP$$

Figure 18-21   Definition of Processes P_{406}, P_{407}, P_{408}, and P_{409}

The operation-based infinite-queue SBC process of the *Bicycle*, A_{401}, is defined as "$!P_{401}\|!P_{404}\|!P_{406}$".

Chapter 19: Operation-Based Infinite-Queue SBC Process of the Restaurant

In this chapter, we use the operation-based infinite-queue SBC process algebra to model the *Restaurant* as shown in Figure 19-1.

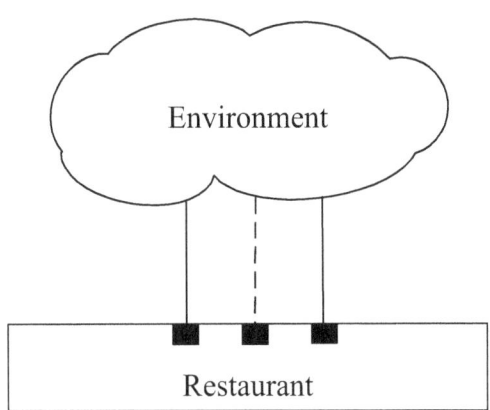

Figure 19-1   Systems Modeling the *Restaurant*

19-1 BNF Tree of the Restaurant

The operation-based infinite-queue SBC process of the *Restaurant*, A_{501}, is defined as "$!(t_{501} \bullet (t_{502}+t_{503}+t_{504}) \bullet (t_{505}+t_{506}+t_{507}) \bullet t_{508} \bullet STOP) || !(t_{509} \bullet STOP)$".

We draw the operation-based infinite-queue SBC process algebra Backus-Naur Form tree of A_{501} as shown in Figure 19-2.

182

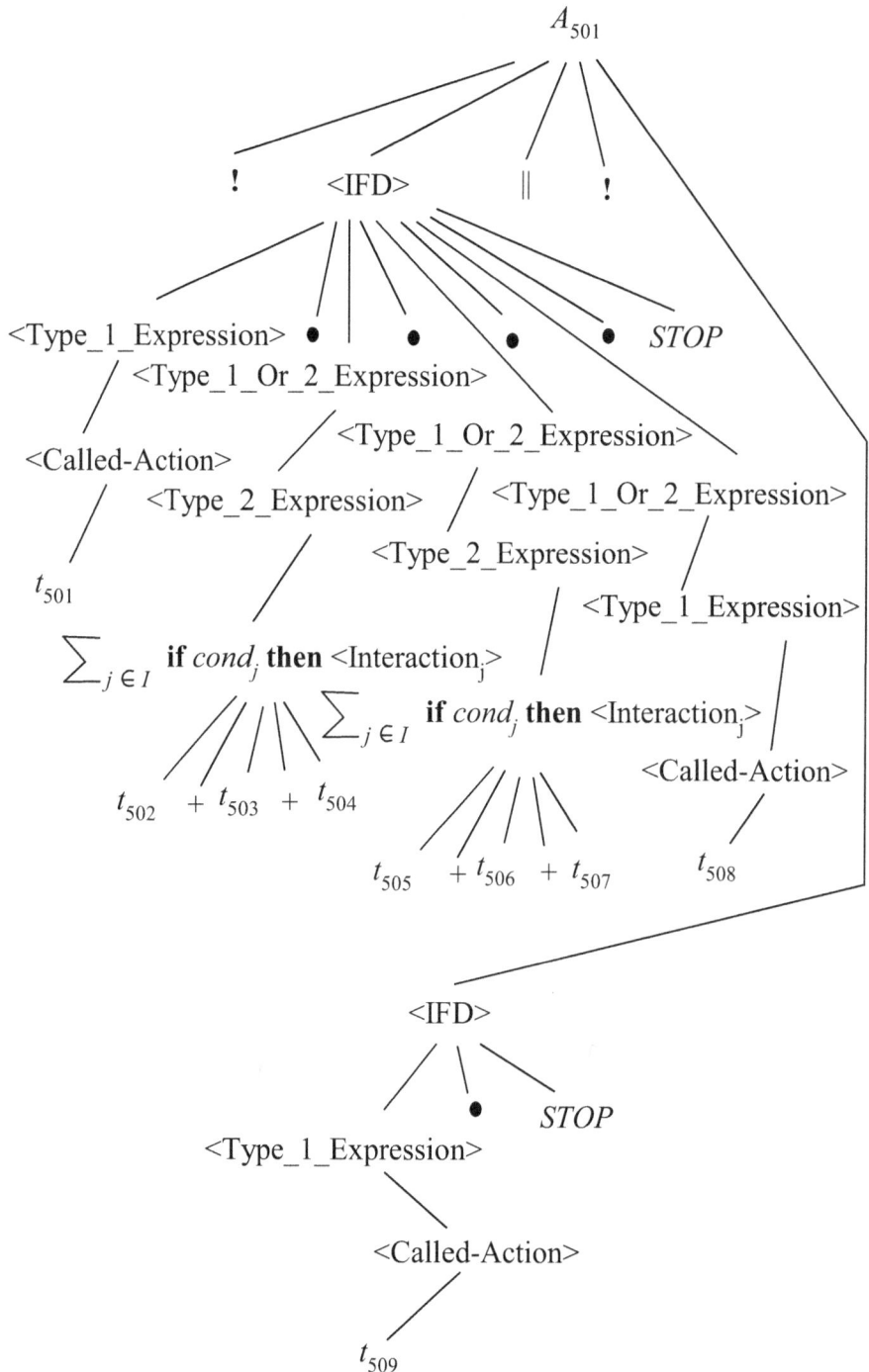

Figure 19-2   Backus-Naur Form Tree of the *Restaurant*'s
Operation-Based Infinite-Queue SBC Process

183

There are two IFDs in the operation-based infinite-queue SBC process of the *Restaurant*. The first IFD is shown in Figure 19-3.

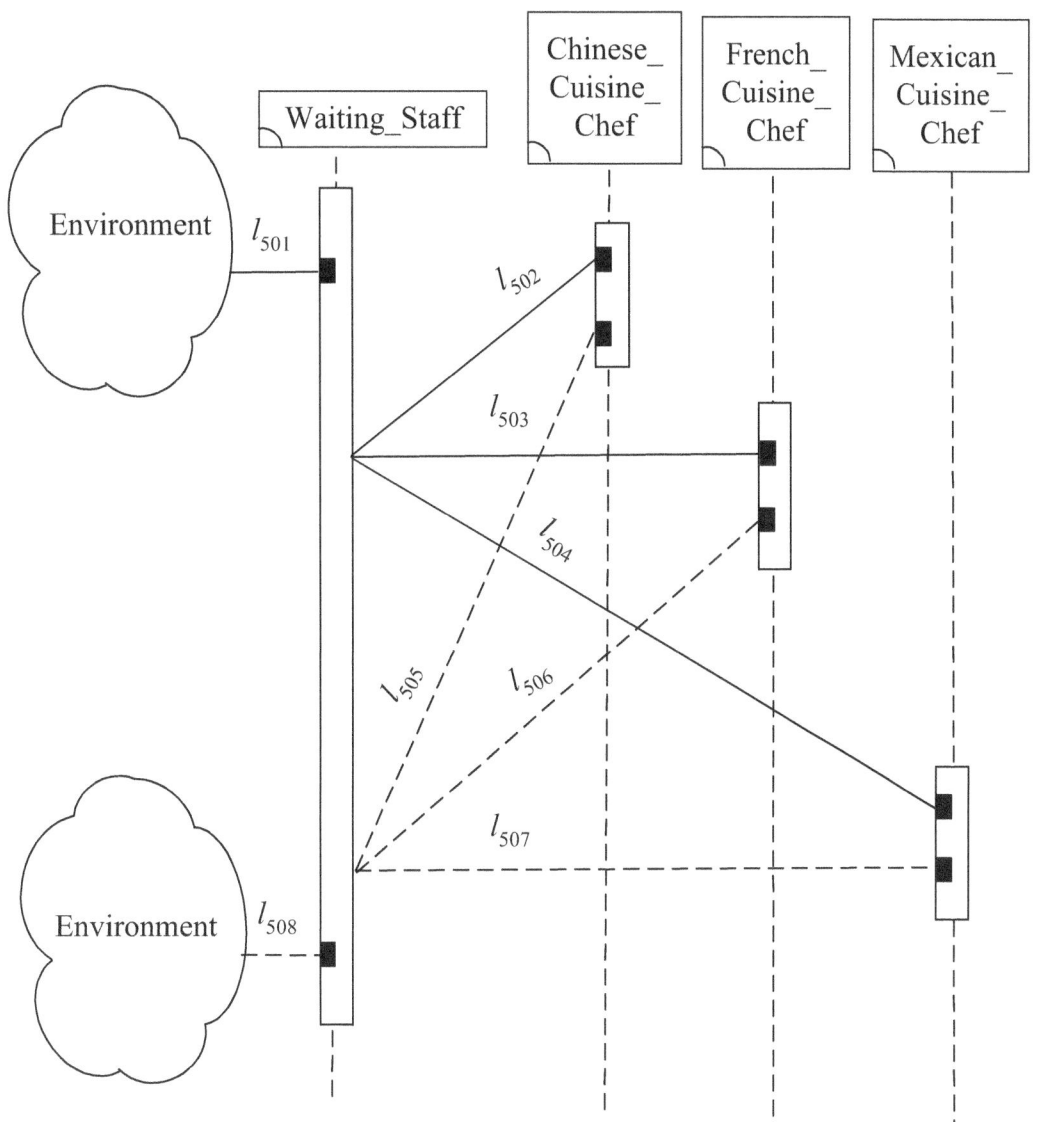

Figure 19-3   First IFD of the *Restaurant*

The second IFD of the operation-based infinite-queue SBC process of the *Restaurant* is shown in Figure 19-4.

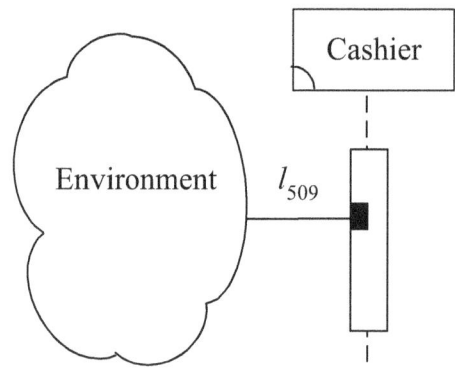

Figure 19-4   Second IFD of the *Restaurant*

19-2 Prefixes of the Restaurant

The t_{501} (operation-based called action with no conditions and a variable setting) prefix is defined as "<OPERATION_CALL, *Waiting_Staff*, CALLED, l_{501}> Order:=Customer_Request", as shown in Figure 19-5.

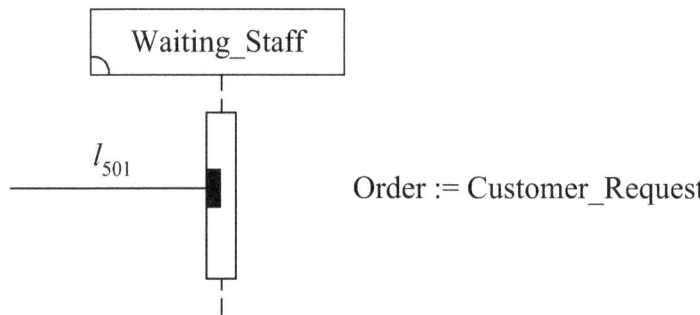

Figure 19-5   Prefix of t_{501}

The t_{502} (operation-based interaction with a condition and no variable settings) prefix is defined as "**if** Order = "Chinese" **then** <OPERATION_CALL, *Waiting_Staff*, l_{502}, *Chinese_Cuisine_Chef*>" as shown in Figure 19-6.

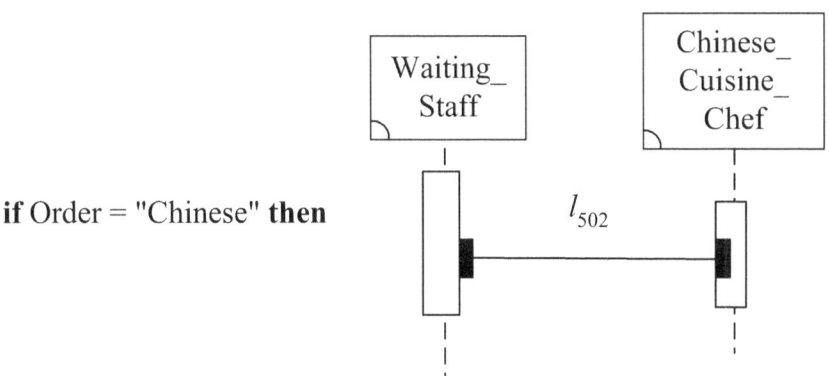

Figure 19-6   Prefix of t_{502}

The t_{503} (operation-based interaction with a condition and no variable settings) prefix is defined as "**if** Order = "French" **then** <OPERATION_CALL, *Waiting_Staff*, l_{503}, *French_Cuisine_Chef*>" as shown in Figure 19-7.

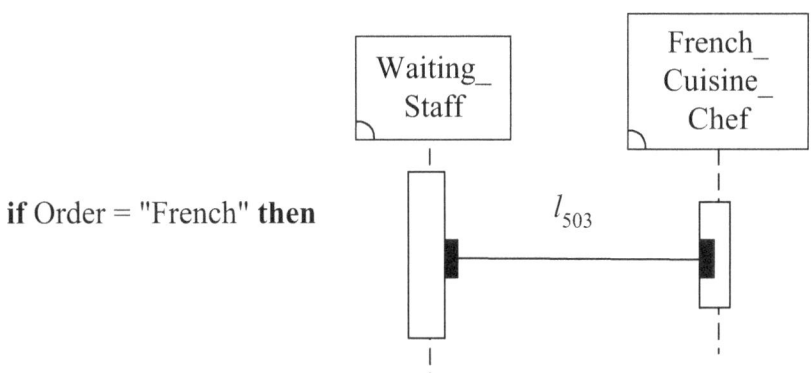

Figure 19-7   Prefix of t_{503}

The t_{504} (operation-based interaction with a condition and no variable settings) prefix is defined as "**if** Order = "Mexican" **then** <OPERATION_CALL, *Waiting_Staff*, l_{504}, *Mexican_Cuisine_Chef>*" as shown in Figure 19-8.

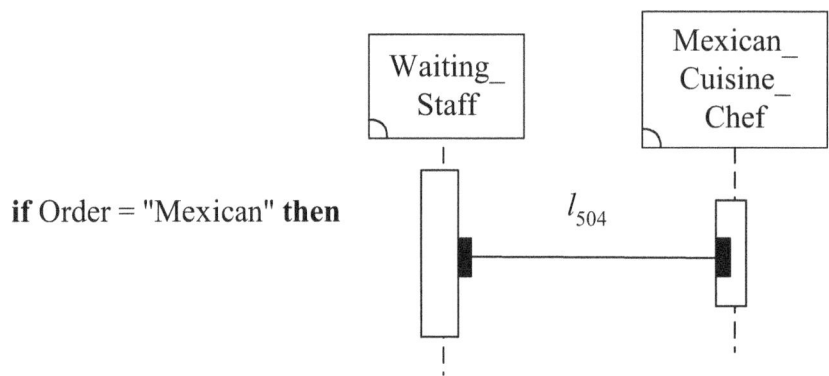

Figure 19-8   Prefix of t_{504}

The t_{505} (operation-based interaction with a condition and no variable settings) prefix is defined as "**if** Order = "Chinese" **then** <OPERATION_RETURN, *Waiting_Staff*, l_{505}, *Chinese_Cuisine_Chef>*" as shown in Figure 19-9.

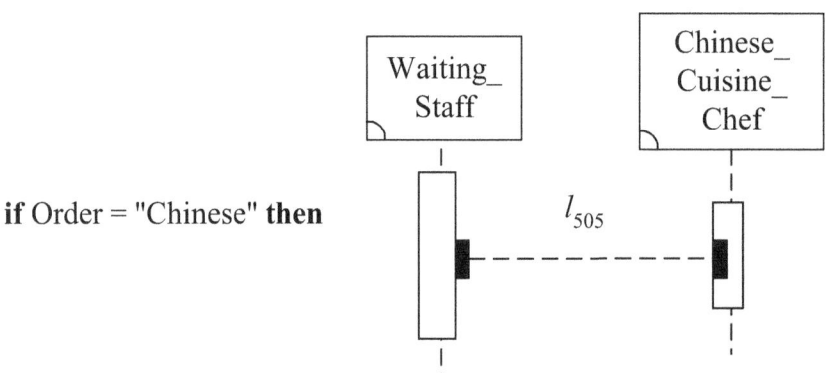

Figure 19-9   Prefix of t_{505}

The t_{506} (operation-based interaction with a condition and no variable settings) prefix is defined as "**if** Order = "French" **then** <OPERATION_RETURN, *Waiting_Staff*, l_{506}, *French_Cuisine_Chef*>" as shown in Figure 19-10.

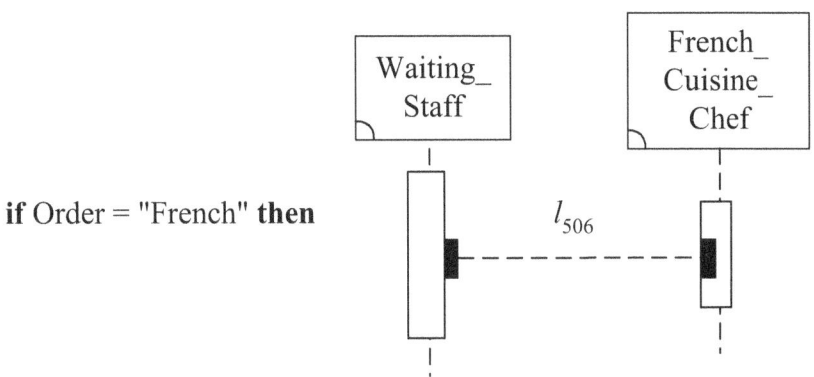

Figure 19-10   Prefix of t_{506}

The t_{507} (operation-based interaction with a condition and no variable settings) prefix is defined as "**if** Order = "Mexican" **then** <OPERATION_RETURN, *Waiting_Staff*, l_{507}, *Mexican_Cuisine_Chef*>" as shown in Figure 19-11.

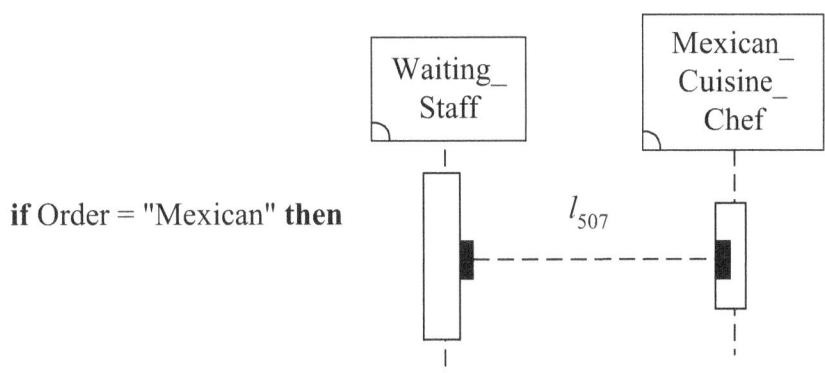

Figure 19-11   Prefix of t_{507}

The t_{508} (operation-based called action with no conditions and no variable settings) prefix is defined as "<OPERATION_RETURN, *Waiting_Staff*, CALLED, l_{508}>", as shown in Figure 19-12.

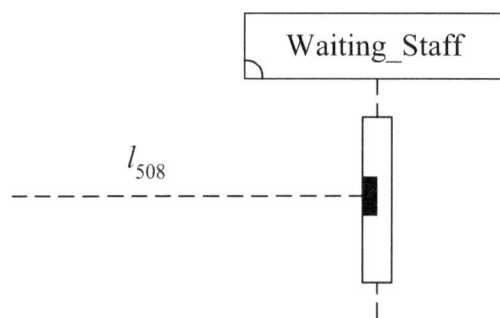

Figure 19-12   Prefix of t_{508}

The t_{509} (operation-based called action with no conditions and no variable settings) prefix is defined as "<OPERATION_CALL, *Cashier*, CALLED, l_{509}>", as shown in Figure 19-13.

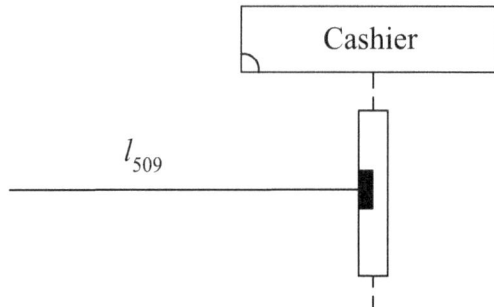

Figure 19-13   Prefix of t_{509}

Figure 19-14 shows all operation call or operation return formulas of the operation-based infinite-queue SBC process of the *Restaurant.*

Entityname	Operation Call Formula	Operation Return Formula
l_{501}	Take_Order(In Customer_Request)	
l_{502}	Cook_Chinese_Food	
l_{503}	Cook_French_Food	
l_{504}	Cook_Mexican_Food	
l_{505}		Cook_Chinese_Food(Out Meal_1)
l_{506}		Cook_French_Food(Out Meal_2)
l_{507}		Cook_Mexican_Food(Out Meal_3)
l_{508}		Take_Order(Out Meal)
l_{509}	Pay_Bills	

Figure 19-14   Operation Call or Operation Return Formulas of
the *Restaurant*'s Process

Figure 19-15 shows the primitive data type specification of the *Order* variable, the *Customer_Request* input parameter, and the *Meal, Meal_1, Meal_2, Meal_3* output parameters.

Parameter	Data Type	Instances
Order	Enumerated	Chinese, French, Mexican
Customer_ Request	Enumerated	Chinese, French, Mexican
Meal	Enumerated food	Chinese food, French food, Mexican food
Meal_1	Food	Chinese food
Meal_2	Food	French food
Meal_3	Food	Mexican food

Figure 19-15   Primitive Data Type Specification

19-3 Process of the Restaurant

The following transition graph shows, in Figure 19-16, the semantics of P_{501}'s operation-based infinite-queue SBC process.

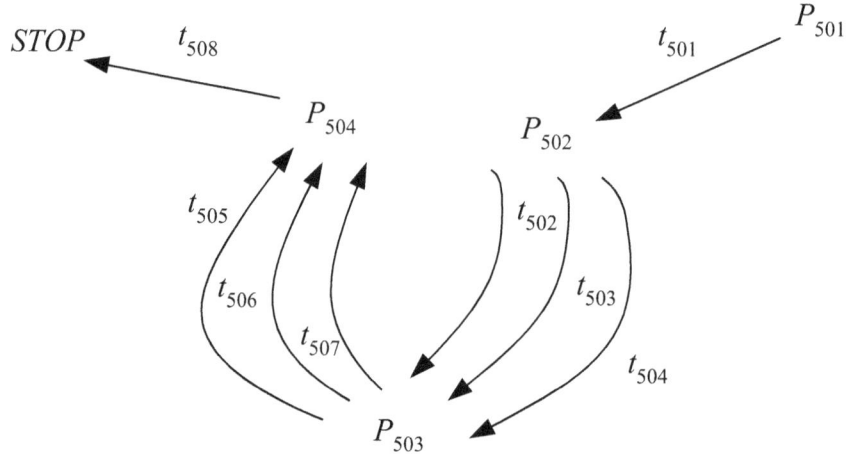

Figure 19-16   Transition graph of the P_{501} Process

In the transition graph of the P_{501}'s operation-based infinite-queue SBC process, processes P_{501}, P_{502}, P_{503} and P_{504} are defined as in Figure 19-17.

$$P_{501} \overset{\text{def}}{=} t_{501} \bullet P_{502}$$

$$P_{502} \overset{\text{def}}{=} t_{502} \bullet P_{503} + t_{503} \bullet P_{503} + t_{504} \bullet P_{503}$$

$$P_{503} \overset{\text{def}}{=} t_{505} \bullet P_{504} + t_{506} \bullet P_{504} + t_{507} \bullet P_{504}$$

$$P_{504} \overset{\text{def}}{=} t_{508} \bullet STOP$$

Figure 19-17   Definition of Processes P_{501}, P_{502}, P_{503}, and P_{504}

The following transition graph shows, in Figure 19-18, the semantics of P_{505}'s operation-based infinite-queue SBC process.

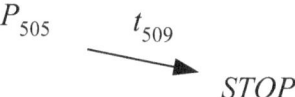

Figure 19-18   Transition graph of the P_{505} Process

In the transition graph of the P_{505}'s operation-based infinite-queue SBC process, process P_{505} is defined as in Figure 19-20.

$$P_{505} \overset{\text{def}}{=} t_{509} \bullet STOP$$

Figure 19-19   Definition of Process P_{505}

The operation-based infinite-queue SBC process of the *Restaurant*, A_{501}, is defined as "$!P_{501}\|!P_{505}$".

193

Chapter 20: Operation-Based Infinite-Queue SBC Process of the Car

In this chapter, we use the operation-based infinite-queue SBC process algebra to model the *Car* as shown in Figure 20-1.

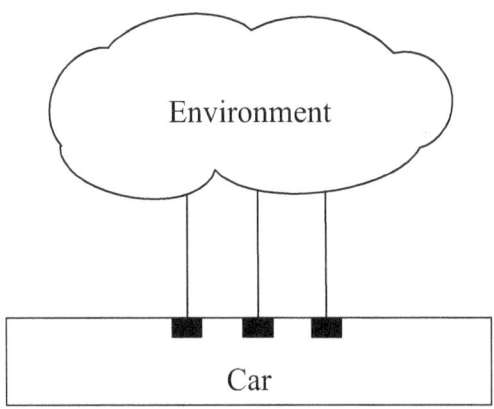

Figure 20-1   Systems Modeling the *Car*

20-1 BNF Tree of the Car

The operation-based infinite-queue SBC process of the *Car*, A_{601}, is defined as "$!(t_{601} \bullet STOP) \| !(t_{602} \bullet STOP) \| !(t_{603} \bullet t_{604} \bullet (t_{605} + t_{606}) \bullet STOP)$".

We draw the operation-based infinite-queue SBC process algebra Backus-Naur Form tree of A_{601} as shown in Figure 20-2.

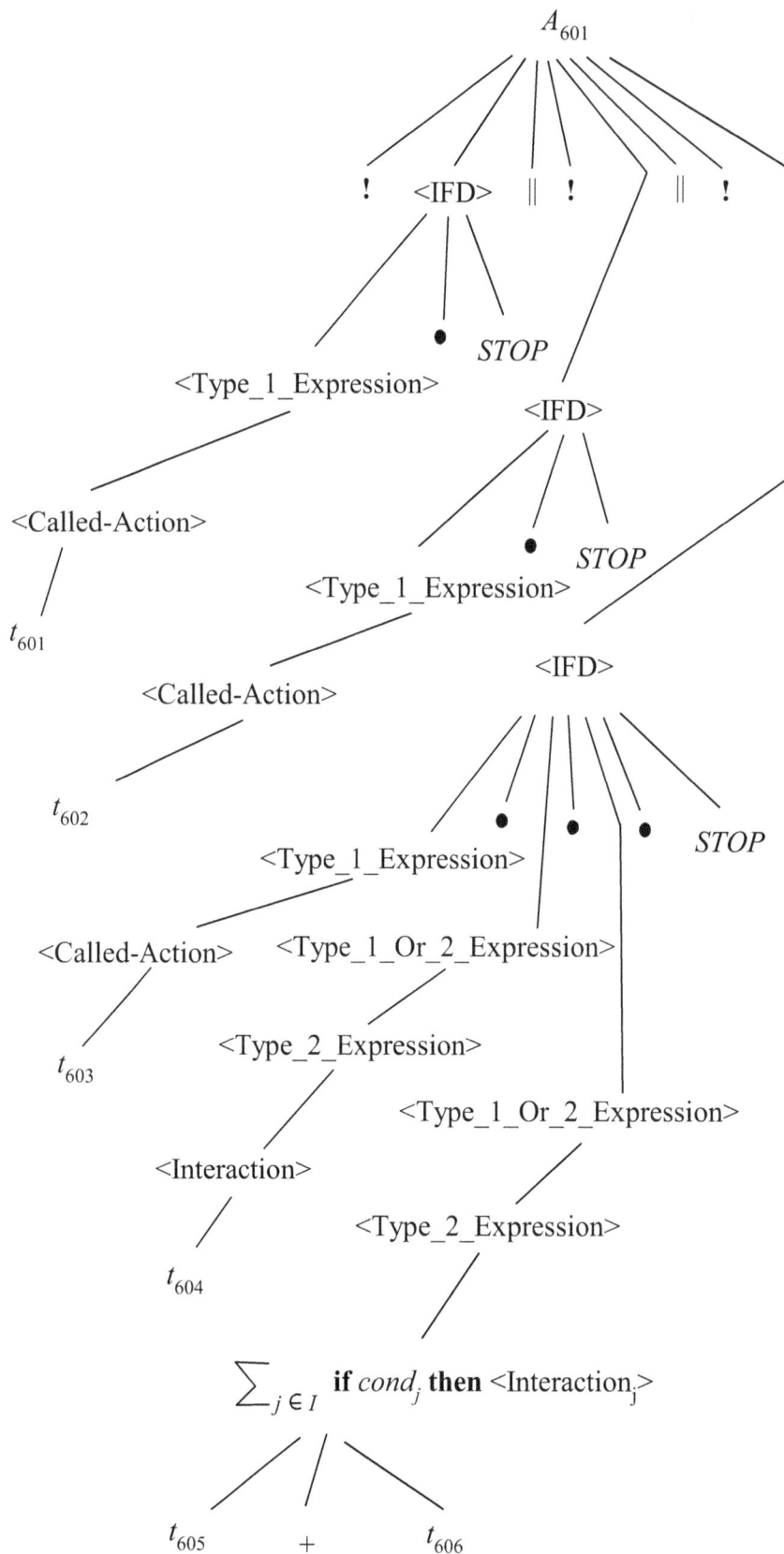

Figure 20-2   Backus-Naur Form Tree of the *Car*'s
Operation-Based Infinite-Queue SBC Process

There are three IFDs in the operation-based infinite-queue SBC process of the *Car*. The first IFD is shown in Figure 20-3.

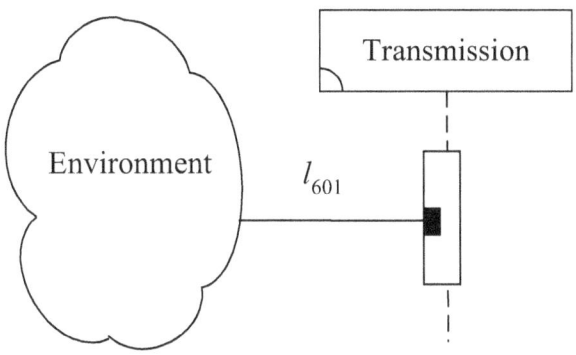

Figure 20-3   First IFD of the *Car*

The second IFD of the operation-based infinite-queue SBC process of the *Car* is shown in Figure 20-4.

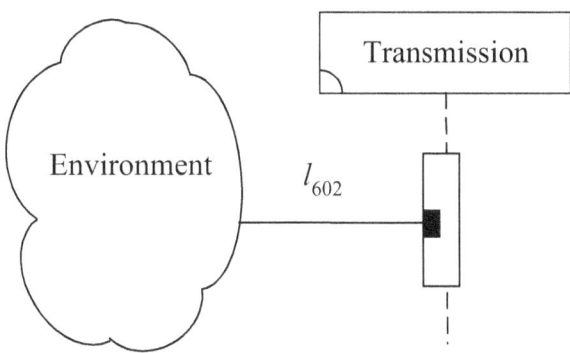

Figure 20-4   Second IFD of the *Car*

The third IFD of the operation-based infinite-queue SBC process of the *Car* is shown in Figure 20-5.

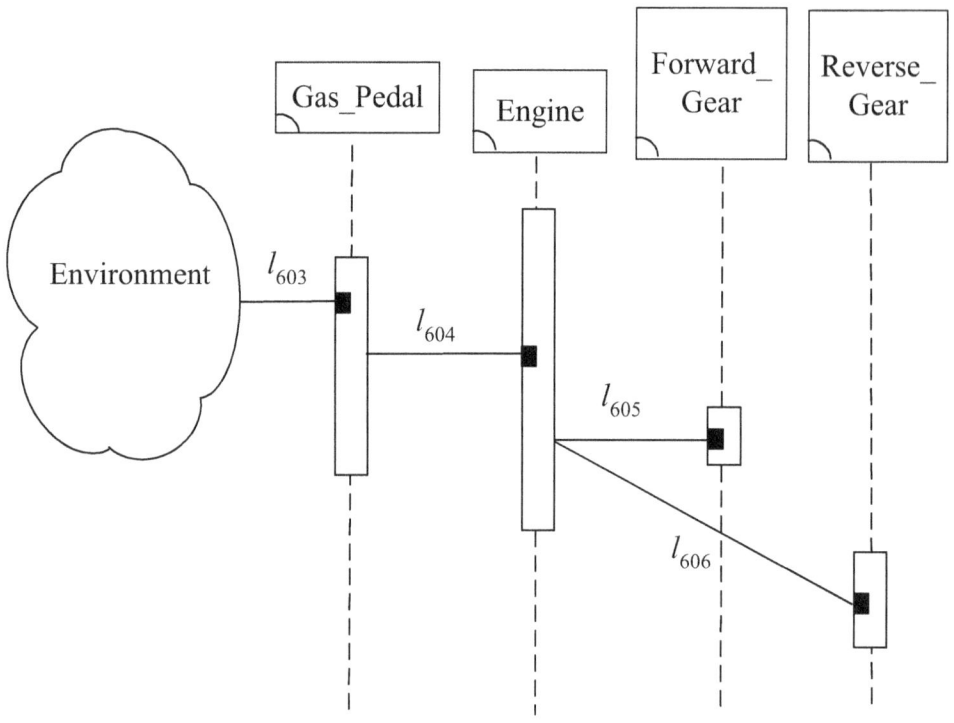

Figure 20-5   Third IFD of the *Car*

20-2 Prefixes of the Car

The t_{601} (operation-based called action with no conditions and a variable setting) prefix is defined as "<OPERATION_CALL, *Transmission*, CALLED, l_{601}> Gear:="Forward"", as shown in Figure 20-6.

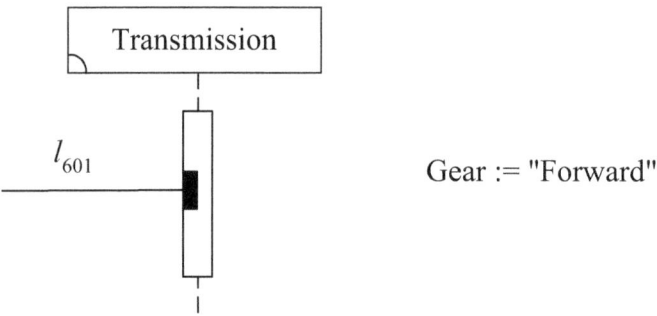

Figure 20-6   Prefix of t_{601}

The t_{602} (operation-based called action with no conditions and a variable setting) prefix is defined as "<OPERATION_CALL, *Transmission*, CALLED, l_{602}> Gear:="Reverse"", as shown in Figure 20-7.

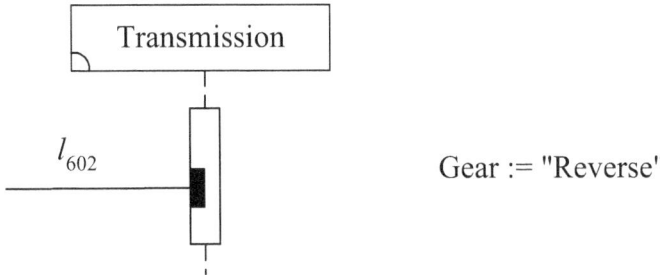

Figure 20-7   Prefix of t_{602}

The t_{603} (operation-based called action with no conditions and no variable settings) prefix is defined as "<OPERATION_CALL, *Gas_Pedal*, CALLED, l_{603}> as shown in Figure 20-8.

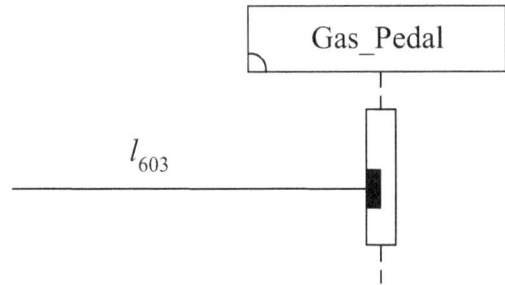

Figure 20-8   Prefix of t_{603}

The t_{604} (operation-based interaction with no conditions and no variable settings) prefix is defined as "<OPERATION_CALL, *Gas_Pedal*, l_{604}, *Engine*>", as shown in Figure 20-9.

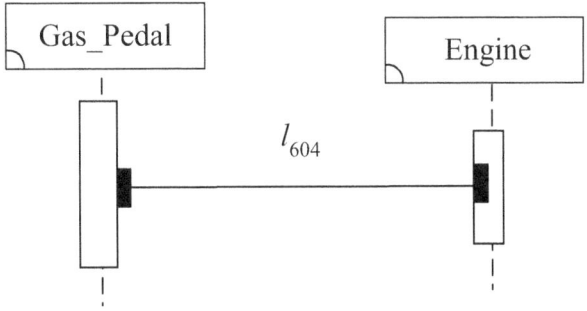

Figure 20-9   Prefix of t_{604}

The t_{605} (operation-based interaction with a condition and no variable settings) prefix is defined as "**if** Gear = "Forward" **then** <OPERATION_CALL, *Engine*, l_{605}, *Forward_Gear*>" as shown in Figure 20-10.

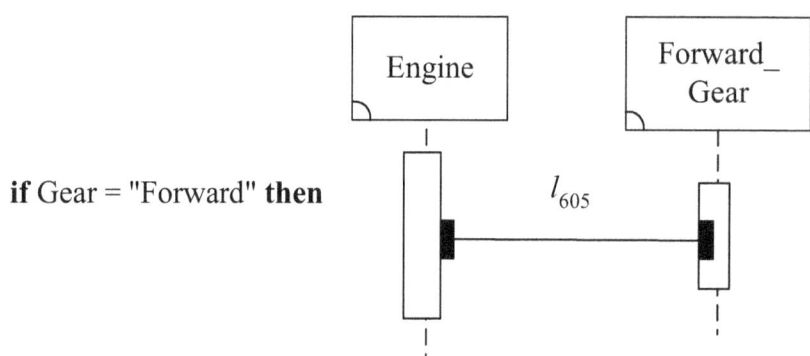

Figure 20-10   Prefix of t_{605}

The t_{606} (operation-based interaction under a certain condition) prefix is defined as "**if** Gear = "Reverse" **then** <OPERATION_CALL, *Engine*, l_{606}, *Reverse_Gear*>" as shown in Figure 20-11.

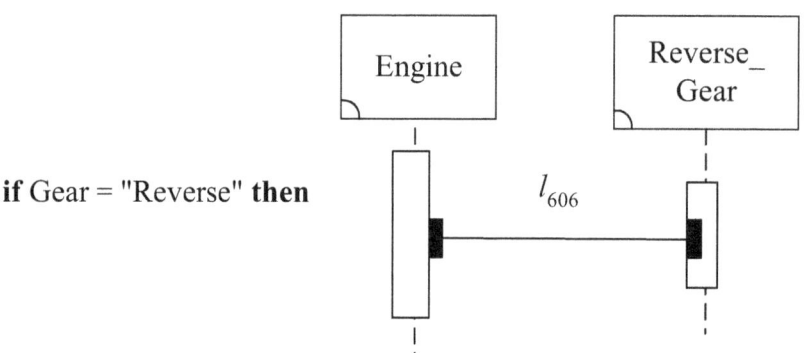

Figure 20-11   Prefix of t_{606}

Figure 20-12 shows all operation call or operation return formulas of the operation-based infinite-queue SBC process of the *Car*.

Entity name	Operation Call Formula	Operation Return Formula
l_{601}	Push_Gear_Forward	
l_{602}	Push_Gear_Reverse	
l_{603}	Depress_Gas_Pedal	
l_{604}	Fuel_Supply	
l_{605}	Forward_Gear_Rotate	
l_{606}	Reverse_Gear_Rotate	

Figure 20-12 Operation Call or Operation Return Formulas of
the *Car*'s Process

Figure 20-13 shows the primitive data type specification of the *Gear* variable.

Parameter	Data Type	Instances
Gear	Enumerated	Forward, Reverse

Figure 20-13   Primitive Data Type Specification

20-3 Process of the Car

The following transition graph shows, in Figure 20-14, the semantics of P_{601}'s operation-based infinite-queue SBC process.

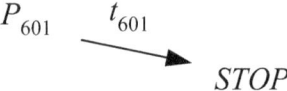

Figure 20-14　Transition graph of the P_{601} Process

In the transition graph of the P_{601}'s operation-based infinite-queue SBC process, process P_{601} is defined as in Figure 20-15.

$$P_{601} \overset{\text{def}}{=\!=} t_{601} \bullet STOP$$

Figure 20-15　Definition of Process P_{601}

The following transition graph shows, in Figure 20-16, the semantics of P_{602}'s operation-based infinite-queue SBC process.

Figure 20-16　Transition graph of the P_{602} Process

In the transition graph of the P_{602}'s operation-based infinite-queue SBC process, process P_{602} is defined as in Figure 20-17.

$$P_{602} \overset{\text{def}}{=\joinrel=} t_{602} \bullet STOP$$

Figure 20-17   Definition of Process P_{602}

The following transition graph shows, in Figure 20-18, the semantics of P_{603}'s operation-based infinite-queue SBC process.

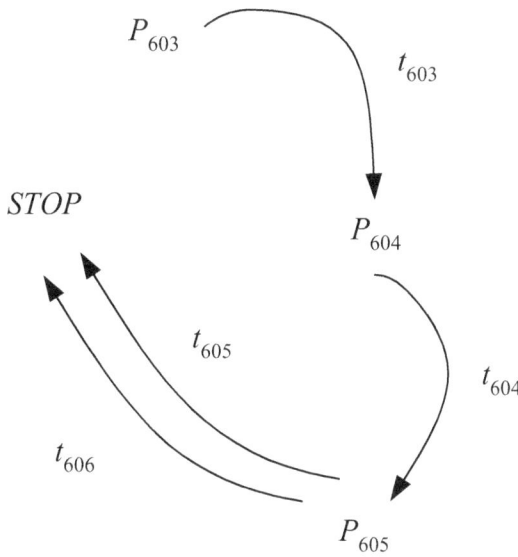

Figure 20-18   Transition graph of the P_{603} Process

In the transition graph of the P_{603}'s operation-based infinite-queue SBC process, processes P_{603}, P_{604} and P_{605} are defined as in Figure 20-19.

$$P_{603} \overset{\text{def}}{=\!=} t_{603} \bullet P_{604}$$

$$P_{604} \overset{\text{def}}{=\!=} t_{604} \bullet P_{605}$$

$$P_{605} \overset{\text{def}}{=\!=} t_{605} \bullet STOP + t_{606} \bullet STOP$$

Figure 20-19   Definition of Processes P_{603}, P_{604}, and P_{605}

The operation-based multi-queue SBC process of the *Car*, A_{601}, is defined as "$!P_{601}\|!P_{602}\|!P_{603}$".

APPENDIX A: Language Constructs of Operation-Based Infinite-Queue SBC Process Algebra

(1) E ::= "**!** "<IFD> {"∥" "**!** " <IFD>}

(2) <IFD> ::= <Type_1_Expression>
{"●" <Type_1_Or_2_Expression>} " ● " *STOP*

(3) <Type_1_Or_2_Expression> ::= <Type_1_Expression>

| <Type_2_Expression>

(4) <Type_1_Expression> ::= <Called-Action>[variable settings]

$$\left| \quad \sum\nolimits_{j \in I} \textbf{if } cond_j \textbf{ then } \text{<Called-Action}_j\text{>[variable settings}_j\text{]} \right.$$

(5) <Type_2_Expression> ::= <Interaction>[variable settings]

$$\left| \quad \sum\nolimits_{j \in I} \textbf{if } cond_j \textbf{ then } \text{<Interaction}_j\text{>[variable settings}_j\text{]} \right.$$

APPENDIX B: Transitional Semantics of Operation-Based Infinite-Queue SBC Process Algebra

Prefix
$$\frac{\rule{3cm}{0.4pt}}{t \bullet E \xrightarrow{t} E}$$

Sum$_j$
$$\frac{E_j \xrightarrow{t} E'_j}{\sum_{i \in I} E_i \xrightarrow{t} E'_j} \ (j \in I)$$

Parallel$_1$
$$\frac{E \xrightarrow{t} E'}{E \parallel F \xrightarrow{t} E' \parallel F}$$

Parallel$_2$
$$\frac{F \xrightarrow{t} F'}{E \parallel F \xrightarrow{t} E \parallel F'}$$

Constant
$$\frac{P \xrightarrow{t} P'}{A \xrightarrow{t} P'} \ (A \overset{\text{def}}{=} P)$$

BIBLIOGRAPHY

[Acko68] Ackoff, R., "Toward a System of Systems Concepts," *Modern Systems Research for the Behavioral Scientist: A Sourcebook*, Aldine Publishing Company, 1968.

[Berg87] Bergstra, J. A. et al., "ACPτ: A Universal Axiom System for Process Specification," *CWI Quarterly* 15, 1987, pp. 3-23.

[Burd10] Burd, S. D., *Systems Architecture*, 6th Edition, Cengage Learning, 2010.

[Chao14a] Chao, W. S., *Systems Thingking 2.0: Architectural Thinking Using the SBC Architecture Description Language*, CreateSpace Independent Publishing Platform, 2014.

[Chao14b] Chao, W. S., *General Systems Theory 2.0: General Architectural Theory Using the SBC Architecture*, CreateSpace Independent Publishing Platform, 2014.

[Chao14c] Chao, W. S., *Systems Modeling and Architecting: Structure-Behavior Coalescence for Systems Architecture*, CreateSpace Independent Publishing Platform, 2014.

[Chao15a] Chao, W. S., *Variants of Interaction Flow Diagrams: The Structure-Behavior Coalescence Approach*, CreateSpace Independent Publishing Platform, 2015.

[Chao15b] Chao, W. S., *A Process Algebra For Systems Architecture: The Structure-Behavior Coalescence Approach*, CreateSpace Independent Publishing Platform, 2015.

[Chao15c] Chao, W. S., *An Observation Congruence Model For Systems Architecture: The Structure-Behavior Coalescence Approach*, CreateSpace Independent

Publishing Platform, 2015.

[Chao15d] Chao, W. S., *Variants of SBC Process Algebra: The Structure-Behavior Coalescence Approach*, CreateSpace Independent Publishing Platform, 2015.

[Chao15e] Chao, W. S., *Single-Queue SBC Process Algebra For Systems Architecture: The Structure-Behavior Coalescence Approach*, CreateSpace Independent Publishing Platform, 2015.

[Chao15f] Chao, W. S., *Multi-Queue SBC Process Algebra For Systems Architecture: The Structure-Behavior Coalescence Approach*, CreateSpace Independent Publishing Platform, 2015.

[Chao15g] Chao, W. S., *Single-Queue SBC Observation Congruence Model For Systems Architecture: The Structure-Behavior Coalescence Approach*, CreateSpace Independent Publishing Platform, 2015.

[Chao15h] Chao, W. S., *Multi-Queue SBC Observation Congruence Model For Systems Architecture: The Structure-Behavior Coalescence Approach*, CreateSpace Independent Publishing Platform, 2015.

[Chao16a] Chao, W. S., *System: Contemporary Concept, Definition, and Language*, CreateSpace Independent Publishing Platform, 2016.

[Chao16b] Chao, W. S., *Systems Architecture of Electronic Toll Collection Cloud Applications and Services IoT System*, CreateSpace Independent Publishing Platform, 2016.

[Chao16c] Chao, W. S., *Systems Architecture of Smart Healthcare Cloud Applications and Services IoT System*, CreateSpace Independent Publishing Platform, 2016.

[Chao16d] Chao, W. S., *Systems Architecture of Smart Healthcare Cloud Applications and Services IoT System*, CreateSpace Independent Publishing

Platform, 2016.

[Chao16e] Chao, W. S., *Systems Architecture of Ridesharing Sharing Economy Cloud Applications and Services IoT System*, CreateSpace Independent Publishing Platform, 2016.

[Chao16f] Chao, W. S., *Systems Architecture of Handy Helper Sharing Economy Cloud Applications and Services IoT System*, CreateSpace Independent Publishing Platform, 2016.

[Chao16g] Chao, W. S., *Systems Architecture of Vacation Rental Cleaning Sharing Economy Cloud Applications and Services IoT System*, CreateSpace Independent Publishing Platform, 2016.

[Chec99] Checkland, P., *Systems Thinking, Systems Practice: Includes a 30-Year Retrospective*, 1st Edition, Wiley, 1999.

[Craw15] Crawley, P. et al., *System Architecture: Strategy and Product Development for Complex Systems*, Prentice Hall, 2015.

[Dam06] Dam, S., *DoD Architecture Framework: A Guide to Applying System Engineering to Develop Integrated Executable Architectures*, BookSurge Publishing, 2006.

[Date03] Date, C. J., *An Introduction to Database Systems*, 8th Edition, Addison Wesley, 2003.

[Elma10] Elmasri, R., *Fundamentals of Database Systems*, 6th Edition, Addison Wesley, 2010.

[Frie11] Friedenthal, S., et al., *A Practical Guide to SysML: The Systems Modeling Language*, Morgan Kaufmann, 2nd Edition, 2011.

[Ghar11] Gharajedaghi, J., *Systems Thinking: Managing Chaos and Complexity: A Platform for Designing Business Architecture*, Morgan Kaufmann, 2011.

[Hoar85] Hoare, C. A. R., *Communicating Sequential Processes*, Prentice-Hall, 1985.

[Maie09] Maier, M. W., *The Art of Systems Architecting*, 3rd Edition, CRC Press, 2009.

[Mead08] Meadows, D. H., *Thinking in Systems: A Primer*, Chelsea Green Publishing, 2008.

[Miln89] Milner, R., *Communication and Concurrency*, Prentice-Hall, 1989.

[Miln99] Milner, R., *Communicating and Mobile Systems: the π-Calculus*, 1st Edition, Cambridge University Press, 1999.

[O'Rou03] O'Rourke, C. et al, *Enterprise Architecture Using the Zachman Framework*, 1st Edition, Course Technology, 2003.

[Putm00] Putman, J. R. et al., *Architecting with RM-ODP*, Prentice-Hall, 2000.

[Rayn09] Raynard, B., *TOGAF The Open Group Architecture Framework 100 Success Secrets*, Emereo Pty Ltd, 2009.

[Roza11] Rozanski, N. et al., *Software Systems Architecture: Working With Stakeholders Using Viewpoints and Perspectives*, 2nd Edition, Addison-Wesley Professional, 2011.

[Scho10] Scholl, C., *Functional Decomposition with Applications to FPGA Synthesis*, Springer, 2010.

[Toga08] The Open Group, *TOGAF Version 9 - A Manual (TOGAF Series)*, 9th Edition, Van Haren Publishing, 2008.

INDEX

www.ingramcontent.com/pod-product-compliance
Lightning Source LLC
Chambersburg PA
CBHW080008210526
45170CB00015B/1924

* 9 7 8 1 5 4 0 7 7 6 7 1 6 *